A Brief History of
Astronomy and Astrophysics

A Brief History of

Astronomy and Astrophysics

Kenneth R Lang

Professor Emeritus, Tufts University, USA

World Scientific

NEW JERSEY · LONDON · SINGAPORE · BEIJING · SHANGHAI · HONG KONG · TAIPEI · CHENNAI · TOKYO

Published by

World Scientific Publishing Co. Pte. Ltd.

5 Toh Tuck Link, Singapore 596224

USA office: 27 Warren Street, Suite 401-402, Hackensack, NJ 07601

UK office: 57 Shelton Street, Covent Garden, London WC2H 9HE

British Library Cataloguing-in-Publication Data
A catalogue record for this book is available from the British Library.

ISBN 978-981-3233-83-6
ISBN 978-981-3235-19-9 (pbk)

For any available supplementary material, please visit
https://www.worldscientific.com/worldscibooks/10.1142/10814#t=suppl

Desk Editor: Christopher Teo

Typeset by Stallion Press
Email: enquiries@stallionpress.com

Dedicated to my grandchildren
Max, Mia, Luca and Lili
and
Their Italian grandmother (nonna) Marcella

Contents

Frontispiece: Woman in the morning sun This portrayal of the glowing sunrise by the German artist Caspar David Friedrich conveys wonder and mystery. It also has a transcendental, mystical quality. The painter once compared the "radiating beams of light" in one of his paintings to "the image of the eternal life-giving Father." (Courtesy of Museum Folkwang, Essen.)

Introduction

From the distant past to the modern era, astronomers have always had an insatiable curiosity about the Universe, and an overwhelming desire to find out what it contains, how it is put together, and how it functions. They have used new telescopes and novel instruments to extend our vision to places that cannot be seen with the unaided eye, discovered a host of unanticipated objects, found out how various parts of the night sky are related, and discovered that the Universe is larger, more complex, and older than has been previously thought. *A Brief History of Astronomy and Astrophysics* traces out the unfolding history of these discoveries, and anchors our present understanding of the Universe within the findings and personalities of accomplished astronomers.

This ongoing search of the unknown began more than four hundred years ago, when Galileo Galilei turned his spyglass to the skies and discovered Jupiter's four large moons. As larger visible-light telescopes were constructed, innumerable unseen stars were found, their motions were deciphered, and the expanding Universe of galaxies was discovered. Only the nearest few of these galaxies can be seen without a telescope to collect their light.

Telescopes that operate at radio wavelengths were next used to discover remote radio galaxies, which can be viewed back to the early stages of our Universe. Observations of those radio galaxies that pass near the Sun led to the accidental sighting of radio pulsars, and the subsequent detection of the binary neutron star that was used to demonstrate the existence of gravitational radiation. Just recently, the sound and waveforms of these rippling gravitational waves have been detected, and attributed to the merger of two black holes into a bigger one on several occasions. And in the meantime, the three-degree cosmic microwave background radiation was unexpectedly discovered while observing radio galaxies during one of the first tests of ground links to a communication satellite.

A Brief History of Astronomy and Astrophysics discusses pervasive movement and relentless change throughout the Cosmos, from its birth to its ultimate fate, and everything in between. These ever-widening horizons of astronomy and our changing perceptions of the Universe were often entirely unanticipated and subjected to controversy, doubt, and even ridicule. They illustrate that the known Universe, which can be observed at any given time, is just a modest part of a much vaster one that remains to be found, often in the least expected ways.

As each day ends, we might watch the Sun set, or notice the Moon move across the sky. Satellites are now employed to glance down at the spinning Earth and up to watch the moving stars or look out at the Universe flying apart. Every planet, star, and galaxy has to remain in movement just to stay suspended in space.

Stars are slowly transforming themselves into new forms as they burn their substance away, and entire galaxies evolve over cosmic time intervals. Nothing stays the same anywhere in the Universe. As we watch everything change, we see how one part affects another and naturally wonder how it all began and the way it might end.

This book includes fascinating biographical details that bring human interest to our emerging understanding of the Universe. Important discoveries have been made by astronomers who came from all walks of life and grew up in diverse regions of the globe, from rural farms to big cities. Some were born into poverty and did not go beyond the early grades in formal education. Others had a privileged youth and studied at the world's most prestigious universities. They have all helped part a cosmic veil to reveal fantastic, previously unknown and wonderfully beautiful aspects of the world.

Accomplished astronomers also partook in the hopes, fears, loves, friendships, mistakes, and enlightenments that the rest of us share. Some of them were driven by ambition and a few were careless about crediting the work of others. Several were humble, quiet, and self-effacing.

Spiritual yearning played an important role in the inquiry, persistence, and discoveries of men who provided the foundation for our celestial science. Many great astronomers, like Johannes Kepler, Galileo Galilei, Isaac Newton, Arthur Stanley Eddington, Henry Norris Russell, and Albert Einstein, participated in a search for an underlying order and pattern to the observable Universe, which they believed had been put in place by its Creator.

Other accomplished astronomers or astrophysicists were either atheists or agnostics, and thought that faith in God was not important to scientific inquiry. Nonetheless, some of them, such as Subrahmanyan Chandrasekhar, believed that their religion provides a rational, tolerant way to live life.

Moreover, most contemporary astronomers and astrophysicists, from Einstein to the present day, have retained belief in an order within the known and vast unknown Cosmos, however it may have originated. This pattern often includes a well-defined predictive behavior that they are now spending their lives trying to find and test.

Our book also helps convey the sense of awe and wonder that we can all experience when looking out at the stars at night. It is as if we are observing, and participating in, something grander and more inclusive than ourselves. On these occasions, Nature becomes a source of joy, even with a sacred depth, and we can marvel at the splendor of the Universe that surrounds us. It can fire our curiosity, help propel us through life, and keep a sparkle in our eyes. We can all regard it with awe, mystery, wonder, and gratitude.

Kenneth R. Lang
Professor Emeritus
Tufts University

Part I
Everything Moves

1. The Earth Moves

"Hereafter, when they come to model Heaven
And calculate the Stars [planets], how they will wield
The mighty frame, how build, unbuild, contrive
To save appearances, how gird the Sphere
With Centric and Eccentric scribbled over,
Cycle and Epicycle, Orb in Orb."

John Milton (1667)[1]

A Universe that can be Counted on

Ancient astronomers must have looked out at the bright beacons of Mars and Venus with a sense of wonder and awe. These celestial vagabonds did not move with the stars. They crossed the sky in a regular pattern that might be used to predict when and where they would next appear. Our ancestors called them *planetes*, the ancient Greek word for "wanderers." The ordered motion of the planets was surely set in place by some great power,[2] and when astronomers tried to describe their wandering movements science began.

As described by the Nobel-prize winning physicist, Robert Millikan, astronomy explains: "A Universe that knows no caprice, a Universe that behaves in a knowable and predictable way, a Universe that can be counted on; in a word, a God who works through law."[3]

The geometrical models that were constructed to describe the planetary movements depended upon one's perspective. About 2000 years ago, the Greco-Egyptian astronomer Claudius Ptolemaeus, provided a complex model involving circular motions in an Earth-centered Universe, which was the best description available for about 1400 years. Then the Polish astronomer Nicolaus Copernicus questioned the Ptolemaic model and set the Earth in motion around the Sun.

3

In the following century, Galileo Galilei brought the Heavens down to Earth by using the newly invented telescope to find otherwise unseen mountains and valleys on the Moon, a star-filled Milky Way, and four moons circling around Jupiter. Galileo's contemporary Johannes Kepler removed the "perfect" circle from consideration of the planetary motions, and replaced it with elongated paths around the Sun. He discovered laws that would provide the foundation of Isaac Newton's theory of universal gravitation.

Pure, Everlasting, Heavenly Music

In antiquity, it was thought that both the planets and stars move in heavenly circles about a central unmoving Earth. Their circular movement had no beginning or end, and would continue forever without change. After all, the Sun and Moon are circular in shape, and wheels move easily across the Earth's ground because they are round.

The ancient Greek philosopher and mathematician Pythagoras of Samos has been credited with the idea that there is music in the spacing of the planets, which emit harmonious sounds related to their distances and speeds of motion. The nearest, slower planets were thought to emit a low sound; the distant faster ones produced a high sound.

Both Plato and Aristotle subsequently developed the concept of the music of the spheres. In his *Republic*, written around 380 BC, Plato advocated circular planetary motions at different uniform speeds in proportion to their distance from the central, spherical Earth.

In his *De Caelo* (*On the Heavens*), Plato's student Aristotle provided a mechanism for the motion by attaching the known planets to seven rotating, crystal-like spheres with a common center, all in counter-rotation to an eighth swift, outermost sphere of stars. Such a stellar sphere would explain why the stars seem to slide across the night sky, and why travelers to new and distant lands see new stars as well as new people.

For Aristotle, the Earth was a place of decay and change, the home of our temporary and impure lives. Natural motions on Earth, as distinguished from forced motions there, travel in straight lines, and that motion always ends. A stone falls straight down and stops, a fire rises straight up and disappears, and every human journey ends. In contrast, the indestructible, pure and eternal planets and stars are in everlasting motion. They seem to last forever and never stop moving.

To the ancient Greeks, the outermost celestial sphere formed the edge of the observable Universe. This sphere contained the fixed stars that remained firmly rooted within the night sky without ever moving with respect to each other. They all moved together as the celestial sphere wheeled around the central Earth once every day. The planets were supposed to move in the opposite direction at a slower pace.

Following Plato's suggestion, astronomers spent centuries trying to describe the observed planetary movements, their appearances, using circular motion at constant speed around a central Earth, but they never could reproduce the temporary backwards motion of Mars, known as a *retrograde*, or its faster and slower motions observed during different parts of its path in the sky.

In the second century AD, the Greek mathematical astronomer Claudius Ptolemaeus created a geometrical model that could "save the appearances" presented by the planets. *Ptolemy*, as he is known, worked in the fine library at Alexandria, Egypt, where he was able to consult the work of previous astronomers. He showed that the observed planetary movements could be described by a system of moving circles in motion around the Earth, like the gears of some fantastic cosmic machine. Each planet was supposed to move with constant speed on a small circle, or epicycle, while the center of the epicycle revolved on a larger circle whose center was displaced from the Earth. A planet in uniform circular motion about a center slightly offset from the Earth would appear to a terrestrial observer to be moving with varying speed, faster when it is closest to Earth and slower when further away.

With this complex arrangement, Ptolemy was able to use circles upon circles to reproduce and predict the apparent motions of the planets with remarkable accuracy. He succeeded so well that his model was still being used to predict the locations of the planets in the sky more than a thousand years after his death.

Then, in the mid-16th century, the Polish cleric Mikolaj Kopernik, better known as Nicolaus Copernicus, set the Earth and other planets moving about a stationary, non-moving Sun.

Copernicus' Vision of Sun-Centered Motion

Nicolaus Copernicus was born on February 19, 1473 in the city of Torún in the Province of Royal Prussia, a region of the Kingdom of Poland. His father was a merchant from the capital Cracow, and his mother, Barbara Watzenrode, was a member of a wealthy and powerful Torún family.

Upon his father's death, when Copernicus was just 10 years old, his mother's brother, Lucas Watzenrode, looked after his education and career. At age 19 he matriculated at the University of Cracow, where his studies included astronomy, mathematics, philosophy, physics, and the works of Aristotle and Ptolemy. Four years later, in 1496, Copernicus began a three-year study of law at the University of Bologna, Italy, a prominent European legal institution. Here he learned Papal decisions regarding authority, judgments, rights, and penalties within the jurisdiction of the Church, known as canon law. To round out his education, Copernicus then began a two-year study of medicine at Padua University in Italy, a leading faculty of medicine.

In 1503, at the age of 30, Copernicus returned home to join the staff of his Uncle Lucas, now the Catholic Bishop of Varmia, which was an area covering about five thousand square kilometers in the far northeast, Baltic coast of Poland, near Gdansk. He spent the next seven years as companion, secretary, and personal physician to his uncle, taking part in administrative, ecclesiastic, economic, and political duties that benefited from his education in canon law.

Before the end of 1510, Copernicus left service with his Bishop-Uncle, and moved to take up duties as the Canon of the Cathedral at Frombork. This town is located at the Vistula Lagoon on the Baltic Sea, far from the centers of European society, and it is in this remote location that the isolated genius would reside for most of his remaining 33 years. His life ended on May 24, 1543 at the age of 70, without ever being married or having children.

In Copernicus' day, just about everyone thought that the Sun and stars were eternally wheeling about the immobile Earth, the center of the Universe. It certainly looked like the Sun was moving around the Earth and across the sky, and even in modern times people still say that the Sun rises and sets, to clock our daily rhythm. The distant stars were similarly thought to revolve around the Earth, and that was why they were seen moving across the dark night sky. Copernicus' great insight was to place the Earth and other planets in uniform, circular motion around a central, stationary Sun.[4]

In the opening remarks of his *Commentariolus*, or *Little Commentary*, circulated around 1510, Copernicus stated that the heavenly bodies move with uniform speed in a "perfect" circle, as Plato and Aristotle had proposed. He then examined Ptolemy's widely used planetary theories in which a planet moves on a small circle around a bigger one, but not really at uniform speed around any circle's center. It only appears in uniform motion when viewed from outside the circle's center, at an "equant" point chosen for that purpose.

Although Ptolemy's theory was good enough to predict planetary motions and positions, Copernicus commented that: "A theory of this sort seemed neither sufficiently absolute, or complete enough, nor sufficiently pleasing to the mind," and he therefore sought a more reasonable arrangement of circles that would "explain all the observed irregularities in planetary motion while keeping everything moving uniformly about its proper center, as required by the principle of perfect motion."[5] He accomplished this by setting the Earth free to revolve about the Sun, which became the immovable center of the Universe.

The Earth became just one planet among five others. They all whirled around the Sun, the source of our light and warmth. It was the daily rotation of the Earth that made the Sun apparently move across the daytime sky and the stars swing by at night.

In Copernicus' Sun-centered model, the Earth and the five other planets visible to the unaided eye swung in the same direction, in uniform circular motion with a period of revolution that increased with the planet's distance from the Sun. In order of increasing distance, they are Mercury, Venus, Earth, Mars, Jupiter and Saturn. [The more distant planets Uranus and Neptune were not discovered until 1781 and 1846, respectively, and that required the use of the telescope that was not invented until the early 1600s.]

As Copernicus noticed, the further a planet is from the Sun, the longer it takes the planet to complete a circuit around the Sun. This vision is conveyed in this extract from his *Revolutions*: "In no other way do we find a wonderful commensurability and a sure harmonious connection between the size of the orbit and the planet's period of revolution."[6] [Copernicus used the relative planetary distances from the Sun expressed in terms of the Earth's mean distance, the "common measure" of the Universe, but no one knew its precise value for an additional three centuries.]

How do the Planets Move?

So who was right? Does the Earth move around the Sun, or is the opposite true? There was no definitive observational test at the time.

Both Ptolemy's Earth-centered motion and Copernicus' Sun-centered one provided different explanations for the temporary backwards motion that had been observed for Mars (Fig. 1.1). It apparently looped back in the wrong direction for weeks at a time, seemingly disrupting its uniform progress across the night sky. The planet gradually came to a stop in its eastward motion, moved toward the west, and then turned around again and resumed moving toward the east. Jupiter and Saturn also displayed such a temporary

Fig. 1.1.   **Retrograde loops** This photograph shows the apparent movements of the planets against the background stars. Mars, Jupiter and Saturn appear to stop in their orbits, then reverse direction before continuing on — a phenomenon called retrograde motion by modern astronomers. (Courtesy of Erich Lessing/Magnum.)

reversed motion in the westward "retrograde" direction before continuing on in the eastward "prograde" direction.

In Ptolemy's Earth-centered model, combinations of uniform circular motions explained these looping, retrograde paths of the planets. As mentioned in Copernicus' *Little Commentary*, and emphasized in his *Revolutions*, the apparent backwards motions can instead be explained by the uniform motion of the Earth and other planets at different speeds around the Sun.

In Copernicus' interpretation, planets moving at a slower speed than the Earth would sometimes appear to move ahead of Earth, and sometimes fall behind. During the relatively short time that the Earth overtakes one of these planets, that planet appears to be moving backward. Moreover, this explained why the size of the retrogrades differs for Saturn, Jupiter, and Mars, and one could confidently predict when their apparent motion would come to a halt and turn around, and for how long they would seem to move in the wrong direction.

Sun-centered planetary motion did provide natural explanations for other planetary observations. If the orbit of Venus lies inside that of Earth and closer to the Sun, it would account for the fact that Venus never appears to venture very far ahead or behind the Sun. An orbit around the Sun might also be adopted to explain the observed track of Venus' motion, which twists and turns in the sky and does not appear to move in a circle around the Earth. In the Sun-centered Copernican model, the orbits of Mars, Jupiter and Saturn lie outside that of the Earth, which explains why they are visible throughout the night.

Nevertheless, when it came to describing and predicting observed planetary motions, Copernicus' Sun-centered model was neither more accurate nor simpler than Ptolemy's Earth-centered one. They were both only approximate accounts of observations. Copernicus was forced to adopt multiple circles to explain the apparent variations in the planetary motions, just as Ptolemy did, and the only major simplification of Copernicus' model was that descriptions of the motions of the Sun and stars were no longer needed. Their movement could be eliminated under the assumption that the Earth rotated daily upon its axis.

In their day, both the Ptolemaic and Copernican models had aspects that critics found hard to believe. In Ptolemy's theory the enormous Sun had to travel around the smaller Earth once each day, and the distant stars had to wheel overhead more rapidly than the Sun every single day. Copernicus found an explanation for the apparent motions of both the Sun and stars, by replacing them with the rotation of the Earth, but his theory strained credibility by placing the stars at a considerably greater distance than had previously been assumed.

If the Earth moved in a great orbit around the Sun, then the stars would have to be very much farther away than the width of the Earth's path. Otherwise, the nearby stars would show a slight change in position when observed from opposite sides of our planet's orbit, which had never been seen. The fact that no such change had been recorded was indeed used as an argument against the Sun-centered hypothesis from the time of Aristotle on.

The slight change in the apparent positions of nearby stars remained undetected for three centuries after Copernicus' death. Before 1838 there wasn't any definite proof that the Earth moves around the Sun.[7] And it was not until 1852 that the rotation of the Earth was conclusively demonstrated using Foucault's famous swinging pendulum.[8]

Copernicus never did present a vigorous defense of the Earth's motion around the Sun, and at the time it was thought to be no more than a conjecture or unproven hypothesis that might be convenient for astronomical calculations and predictions. Planetary movement about a central, unmoving Earth continued to be the favored world-view for about a century after Copernicus' death, and it wasn't until around 1700 that his idea became widely accepted. In the meantime, Galileo had used the newly invented telescope to bring the Heavens down to Earth, and the Earth into the Heavens, and in 1633, the Catholic Church had summoned Galileo before a Holy Inquisition for supporting Sun-centered planetary motion.

Heavenly Things Never Seen Before

Galileo Galilei was born in Pisa, Italy in 1564, and died at the age of 77 at his home in Arcetri within the hills surrounding Firenze (Florence), Italy. His long life included acclaim and fame, condemnation and house arrest, and emotional extremes ranging from passion to despair. He was ambitious and gregarious, proud and humble, religious and independent minded, and both humorous and sarcastic.

After years of study and teaching in Pisa, Galileo was appointed by the Venetian Senate to the vacant Chair of Mathematics at the University of

Padua near Venice, where he taught geometry, mechanics and astronomy for 18 years, from 1592 to 1610. As Galileo would later claim, these were the best years of his life, filled with novel ideas, alert colleagues, new friends, and his beautiful Venetian mistress Marina Gamba. The colorful Galileo was also a notorious lover of good wine, which he called "light held together by moisture."[9]

Galileo was already middle aged when the telescope, or *spyglass* as it was first called, was invented.[10] When he heard of the device that brought faraway things nearby, Galileo promptly built improved versions of it and anticipated its military advantages for defense of Venice, a walled city beside the sea.[11]

When Galileo demonstrated his telescope to the Venetian senators, and gave them one as a gift, he was awarded by a substantial increase in salary and granted lifelong tenure at Padua. The inquisitive professor then turned one of his telescopes toward the sky, and revealed things never seen before.[12]

In Galileo's time, many astronomers supposed that the Milky Way was composed of a misty, nebulous substance, but his spyglass showed that it instead consists of a multitude of stars that cannot be seen with the unaided eye. He also turned his telescope to the Moon and was able to resolve features that otherwise remained blurred or unseen. Chains of lofty mountains, deep valleys, and round craters were discovered on the Moon's surface. It is pock-marked, cracked, and molded into high and low places, just like the Earth.

Then in January 1610, the innovative Galileo made another startling discovery that further diminished the central specialness of the Earth. When directing his telescope at the nearly full Moon, Galileo must have naturally moved his spyglass just a little to look at Jupiter, which was then located just above the Moon and the next brightest object in the sky. Perhaps for the first time in history, he observed a planet not as a point, but as a round disk. Moreover, near Jupiter he saw three unresolved objects all in a straight line passing through the planet's disk. The three compan-ions were detected on the next night, and at first he thought they were stars. But one of them disappeared the next two nights, and then reap-peared. On the following night, Galileo for the first time saw four objects near the planet Jupiter (Fig. 1.2).[13]

Galileo became convinced that he was seeing four moons that revolved about Jupiter the way our Moon moves around the Earth. They accompanied

DISCOVERY OBSERVATIONS OF JUPITER'S FOUR LARGE MOONS

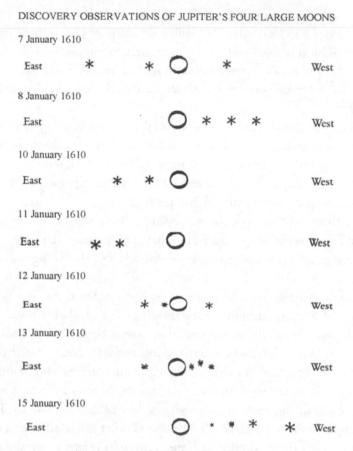

Fig. 1.2. Moons of Jupiter Some of Galileo Galilei's observations of the "Medicean stars", which were drawn in his *Sidereus Nuncius* of 1610. They are lined up on each side of Jupiter and change apparent position while orbiting the planet.

Jupiter in its motion across the sky, and traveled regularly around the planet at different distances and speeds.

No one had predicted the possible existence of moons orbiting any other object than the Earth, and the Jovian moons conclusively showed for the first time that the Earth is not the only center of heavenly motion.

As Galileo subsequently demonstrated, the orbital periods of the Jovian moons increase with their distance from Jupiter, from 42 hours to half a month, which meant that the closer moons systematically move faster around the planet than the more distant ones.

Every possible evening Galileo recorded the positions of Jupiter's moons, while also writing a report of his pioneering discoveries with the telescope and publishing it in March 1610. The short book, *Sidereus Nuncius* or *The Sidereal Messenger*, is one of the most fascinating and lively books in astronomy.[14]

Sidereus Nuncius was written in Latin to make Galileo's results accessible to international scholars, and the result was overwhelming. All of Europe was abuzz with excitement over this treatise about previously unseen features on the Moon and new moons that circle Jupiter. These were incredible discoveries, and anyone who heard about them must have been captivated by the wonder of it all. There were hidden things out there that no one had ever seen before.

Barely ten months after Galileo's book was published, the English poet John Donne captured the essence of the discoveries, writing:

> "Man hath weaved out a net, and this net thrown
> Upon the Heavens, and now they are his own."[15]

Under Galileo's telescopic scrutiny, the Moon, Sun, and planets suddenly became physical objects with irregularities, spots, and moons of their own. They were no longer the "perfect" heavenly jewels imagined by Aristotle. Heaven had been brought down to Earth and the Earth up into Heaven. The Earth could no longer be considered the center of all heavenly motion, for Jupiter had moons that revolved around that planet rather than the Earth. Galileo also opened the skies to vast numbers of stars that could only be seen with a telescope.

The enthusiastic Galileo continued with his amazing telescopic discoveries that included the phases and variations in apparent size of Venus. To preserve his priority before being sure of his findings, Galileo circulated the results as Latin anagrams, or successions of scrambled Latin letters. Within a few months, when he had confirmed the findings, he then sent his correspondents the unscrambled solutions, which when translated into English read: The mother of Loves [Venus] emulates the shapes of Cynthia [the Moon].

Venus goes through a complete sequence of Moon-like phases, varying from a thin crescent to a round disk (Fig. 1.3), which meant that "Venus revolves about the Sun."[16] [In the Ptolemaic system the epicycle of Venus always lay between the Earth and the Sun, so if the planet shined by reflected

FIVE PHASES OF VENUS

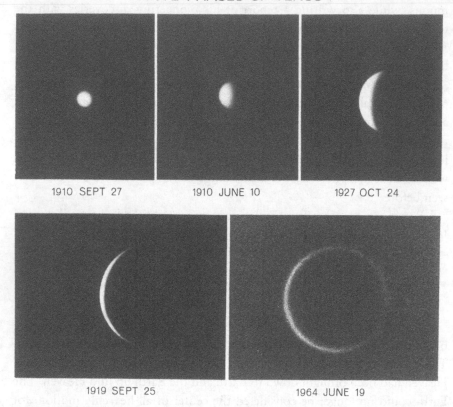

1910 SEPT 27	1910 JUNE 10	1927 OCT 24
1919 SEPT 25		1964 JUNE 19

Fig. 1.3.   Venus The planet Venus changes in both the amount of sunlight it reflects and in its apparent size. (Lowell Observatory photographs.)

sunlight and orbits the Earth it could never show a full phase.] Nevertheless, this did not prove that the Earth revolves around the Sun.

Under scrutiny with his telescope, Galileo also found that Saturn did not always seem to be perfectly round but elongated. When explained and translated, his anagram read: I have observed the highest planet [Saturn] to be triple-bodied. The blurry objects that Galileo saw on each side of Saturn in 1610 disappeared two years later, when Galileo wondered if the planet "had devoured her children." The paradox of Saturn's disappearing appendages wasn't resolved until 1656 when the Dutch astronomer Christiaan Huygens realized that their geometry suggested a narrow ring.

Galileo's telescope also indicated that the apparent perfection of the Sun is an illusion. To most of us, the Sun looks like a faultless, white-hot globe, round, smooth and without a blemish, but detailed scrutiny indicates that dark, ephemeral spots, called *sunspots*, deface the apparently serene face of the Sun. Although Chinese observers had previously noticed the largest sunspots, which can be seen without a telescope, Galileo was one of the first to use a telescope to see smaller sunspots and determine how they came into view, underwent transformations, and disappeared from sight.

The sunspots were always changing shape, and remained visible for hours to weeks before apparently moving back inside the Sun. Altogether, they demonstrated that the Sun was not the unblemished and unchanging heavenly object described by the ancients, but instead: "In that part of the sky which deserves to be considered the most pure and serene of all — I mean in the very face of the Sun, these innumerable multitudes of dense, obscure, and foggy materials are discovered to be produced and dissolved continually in brief periods."[17] By observing a single, long-lived spot, Galileo even demonstrated that the Sun is spinning in space, and turning around once every month or so.

These studies of the motions of moons and planets, as well as his investigations of regular pendulum movements on the Earth, led Galileo to realize the universality of all movements, whether they be in a straight line or rounded into a circular or parabolic trajectory.

Galileo's Lifelong Faith and Inquisition

For centuries, Galileo's disputes with the Roman Catholic Church have symbolized defiance of authority and freedom from religious suppression. Nevertheless, he had an indestructible faith, retained strong belief in God throughout his life, and did not intend to undermine prevailing spiritual beliefs. Galileo remained a good Catholic and was committed to the Church throughout his life. He also never criticized the importance of the *Holy Bible*, just interpretations of some of its passages.

When he discovered new features on the Moon and the four largest moons of Jupiter, in 1610, the devout Galileo wrote to the Tuscan court: "I infinitely render grace to God that it has pleased him to make me alone the first observer of an admirable thing, kept hidden all these ages,"[18] and in

1613, in his third *Letter on Sunspots*, he included: "Whatever the course of our lives, we should receive them as the highest gift from the hand of God…. Indeed, we should accept misfortune not only in thanks, but in infinite gratitude to Providence, which by such means detaches us from an excessive love for earthly things and elevates our minds to the celestial and Divine."[19]

Galileo nevertheless believed that the Earth and everything on it were spinning, swift moving through space, and circling the Sun like all the other planetary wanderers. This thought was downright discomforting to powerful clergy who noted that the *Bible* indicates that the Sun moves and the Earth stands still, not the other way around.

In a letter written to his friend, the monk Benedetto Castelli, and privately circulated in the closing days of 1613, Galileo stated: "Though *Holy Scripture* cannot err, nevertheless some of its interpreters and expositors can sometimes err in various ways … when they would base themselves always on the literal meaning of words."[20] As Galileo pointed out, in the 4th century Saint Augustine had already written that the *Bible* did not need to be interpreted strictly or used to understand the course of the Sun and Moon. To make the point, Galileo quoted in 1615 his contemporary Cardinal Cesare Baronio, who stated that the *Bible* teaches "how to go to Heaven, not how the Heavens go."[21]

After further correspondence, disputes, and intrigue, the Roman Catholic Church decided to take action. In 1616, at the request of Pope Paul V, the cardinals of the Holy Office in Rome examined the Copernican system and found it to be false and contrary to *Holy Scripture*. Cardinal Roberto Bellarmine, a foremost theologian of his day who was subsequently declared a Saint, had already noted that no one had decisively shown that the Earth moves. "To demonstrate that the appearances can be saved by assuming the Sun is at the center," he exclaimed, "is not the same thing as to demonstrate that *in fact* the Sun is at the center and the Earth is in the Heavens."[22] In this the Cardinal was not mistaken, for Galileo had not proved that the Earth moves, and he had not conclusively shown this to be the truth.

There was a temporary change in attitude in 1624, when a new Pope Urban VIII, an admirer of Galileo and his telescope, told him that he saw no harm in his using the Sun-centered system as a tool for astronomical calculations and predictions — even to write about it — as long as he considered it an unproved hypothesis and gave equal treatment to different points of view.

So the tide had apparently turned. Galileo had been given a way out, even felt encouraged, but he refused to compromise, and responded to the friendly gesture with a combative *Dialogo sopra i due massimi sistemi del mondo (Dialogue Concerning the Two Chief World Systems)*, which compared the relative merits of the Earth-centered Ptolemaic system and the Sun-centered Copernican one. In 1630, at the age of 66, Galileo took the finished manuscript to Rome to obtain the approval of the Roman Catholic Church. After some changes suggested by the chief censor and a two-year delay, in part resulting from the spread of the Bubonic Plague into Italy, a thousand printed copies of the *Dialogo* appeared in Firenze, Italy in 1632.

The book was written in Italian with a combative, funny, and at times poetic style, certain to please Galileo's friends and to entertain the general public. But he misjudged its likely reception by the Catholic Clergy, and it didn't help that Galileo adopted a scornful attitude toward those "mental pygmies" who held different views. "Philosophers," he wrote, "fly alone like eagles, and not in flocks like starlings."[23]

It was a definite mistake to put Pope Urban VIII's words — about human ineptitude in understanding a Universe created by an all-powerful God — in the mouth of Simplicio, an apparent simpleton. The Pontiff thought he was being mocked and became an implacable foe. Other opponents thought that the Biblical truth was being threatened, and that the human-centered, Earth-based view of Creation was endangered.

In an oft-told story, the Inquisition summoned Galileo to Rome in 1633. Sick with all manner of afflictions, including gout, arthritis, kidney stones and hernias, and most likely terrified of what might happen to him, the aging and feeble Galileo recanted his physical and astronomical reasons for conclusively supporting the idea that the Earth moved around the Sun.

But the Cardinal-Inquisitors and Pope Urban VIII could show no mercy, partly because other factors were at play. The "Thirty Years" War against the German Protestants was raging through Europe, and the Pope had been openly censured for not defending the Catholic faith. He thought he could support his religion by condemning Galileo for his views.

Galileo was convicted of challenging the authority of the Church, and forced to read a prepared confession that he "abjured, cursed and detested" his erroneous belief that "the Sun is motionless in the center of the world, and the Earth is not the center and moves."[24] The *Dialogo* was permanently banned in the *Index of Prohibited Books*, and after being placed in custody at the palace of the Archbishop of Siena; Galileo was confined to his house in Arcetri, in the hills surrounding Firenze, where he spent his last years. Not until 1992 did Pope John Paul II express regret for how the Galileo affair was handled, and the Catholic Church then declared that theologians of Galileo's time were mistaken because of their literal interpretations of *Sacred Scripture*.

However, Galileo was never imprisoned, as in a prison cell. He continued to correspond with and even receive distinguished visitors, and to write his important *Two New Sciences*. Moreover, his forced confession and house confinement did not shake Galileo's belief in a Creator God. He remained devout until his death.

In the meantime, Tycho Brahe had obtained accurate observations of the shifting locations of planets, and Johannes Kepler used them to show that the planets move with irregular speed along elliptical orbits centered on the Sun. To Kepler, this was in agreement with God's Creation of Divine and harmonious planetary motion that played His heavenly music. As a Protestant outside the jurisdiction of Rome, Kepler avoided religious persecution for publishing these "revolutionary" ideas in 1609 and 1619, well before Galileo's *Dialogo* of 1632.

Kepler's Sacred Mystery and Divine Harmony

Johannes Kepler was born on December 27, 1571 into what was once a prominent Protestant family in the small village of Weil der Stadt in southwest Germany, now part of the Stuttgart region near the Black Forest and the Rhine. [He was therefore just seven years younger than Galileo.]

Johannes had a total of six brothers and sisters. Three died in childhood, two led normal adult lives, and a third Heinrich, was an epileptic misfit. Johannes was himself born premature and almost died from childhood smallpox, which left him with weak vision and crippled hands.

Kepler probably found solace and refuge from such an unhappy beginning in astronomy, mathematics, and religion, which would have drawn his attention away from his everyday life. He was the first in his family to be sent to an elementary school, where he was seen to be an unusual student and was therefore transferred from a German to a Latin school. Since the University of Tübingen was full, Kepler completed his undergraduate work at a preparatory school, and then matriculated at Tübingen with the intent of becoming a Protestant minister.

After qualifying to become a pastor, Kepler took a job as a teacher of mathematics and astronomy at Graz, which is now the second-largest city in Austria after Vienna. While at Graz, Kepler published his *Mysterium cosmographicum*, or *The Sacred Mystery of the Cosmos*, where he interpreted the relative sizes of six planetary orbits around the Sun in terms of six spheres with five geometric solids between them. When nested inside each other they could explain the relative sizes of the planetary orbits, in a geometric design that Kepler believed was created by God.

While at Graz, Kepler married Barbara Müller, who was already twice widowed at the age of twenty-three. In 1597 they embarked on a marriage that was just about as unpleasant to Kepler as his childhood, and ended fourteen-years later when Barbara passed away with a disturbed mind.

Meanwhile, after refusing to convert to Catholicism, Kepler and his family were banished from Graz, and traveled to Prague where he began work with the Danish nobleman Tycho Brahe. With royal patronage, Brahe had built an awesome private observatory, Uraniborg, on the island of Hven, where he spent twenty years amassing the long, exact observations required for a good understanding of planetary motion, including especially precise and detailed measurements of the changing location of Mars. This was before the invention of the telescope, and Tycho used ingenious measuring instruments with graduated circles but without any lenses or mirrors to obtain a then incomparable angular positional accuracy of about one minute of arc.

In 1600 Johannes Kepler began his attempts at explaining Tycho's precise observations of Mars, and kept at the task for nearly 10 years before he could fully account for them. He spent years trying to explain the observations under the assumption that Mars moves in a circle, but the calculations always disagreed with the observations by a frustrating 8 minutes of arc. With extraordinary dedication, endurance and patience, Kepler kept at the task until he found that a non-circular, elliptical orbit could be used to make predictions of Mars' position in the sky accurate to a few minutes of arc. [Shortly after Tycho's unexpected death on October 24, 1601, Kepler was appointed imperial mathematician to the Holy Roman Emperor Rudolph II.]

Kepler apparently reached his startling conclusion that Mars moves at a varying speed along an oval trajectory in 1605, but disagreements with

Tycho's heirs prevented publication of this result until 1609, in *Astronomia nova seu physica coelestis*, the *New Astronomy or Celestial Physics*.[25] It explained the motion of Mars and presented the first two of his now-famous laws of planetary motion. In the first law, Kepler abandoned circular motion, and proposed that all the planets move along an ellipse with the Sun at one of two foci. The term *focus* was used by Kepler to designate "hearth," because to him the Sun was at the hearth and heart of the Universe.

His realization that an elliptical orbit would match the observed path of Mars in the sky resembled an epiphany, which Kepler described: "As if I were roused from a dream and saw a new light."[26] He believed that God had guided him towards the problem and its solution, writing: "I believe it was an act of Divine Providence that I arrived just at the time when Longomontanus [Tycho Brahe's assistant before Kepler] was occupied with Mars. For Mars alone enables us to penetrate the secrets of astronomy which otherwise would remain forever hidden from us."[27]

Because observations indicated that Mars moves a little faster when nearest the Sun than when further away, Kepler proposed his second law, known as the law of areas, that specifies such a variation of speed along the orbit. Assuming that the Sun is the source of all planetary motion, Kepler imagined that an invisible line connected the central Sun to each planet. Remembering that Archimedes found a circle's area by dividing it into a large number of triangles, Kepler imagined that a planet sweeps out triangles as it moves along its elliptical orbit. If triangles with equal areas are swept out in equal time, which is the law of areas, then the planet moves fastest when nearest the Sun (Fig. 1.4).

This dependence of a planet's orbital speed on its distance from the Sun also suggested to Kepler that planetary motion is driven by the Sun, with a motive power that weakens with increasing distance from it. He would eventually show that this weakening applied not only to a single planet along its oval orbit, but also to different planets, with the more distant planets moving about the Sun at slower speeds.

Kepler's patron, Rudolph II, died in January 1612, and early that year Kepler moved to Linz, Austria, on the Danube, where he served as a teacher at the district school. He then enjoyed nearly two decades of financial security and religious freedom. Already in 1613, at the age of 41, Kepler had begun a second, happier marriage to Susanna Reuttinger, aged 24. Two years

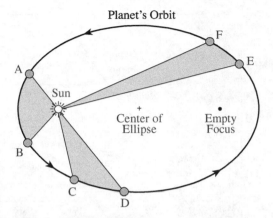

Fig. 1.4. Kepler's first and second laws The orbit of a planet about the Sun is an ellipse with the Sun at one focus, and the line joining a planet to the Sun sweeps out equal areas in equal times. The planet therefore changes speed in its orbit and moves fastest when closest to the Sun.

later, he successfully defended his mother at her trial for witchcraft and consorting with the Devil. [She used herbs to make potions that she believed had magical powers].

In another four years Kepler had written and published his infamous *Harmonices mundi*, or *Harmony of the World*, which extends his investigations of Mars to the orbits of the other known planets. It describes what is now known as his third law of planetary motion, in which the squares of the periods of revolution of any two planets about the Sun are proportional to the cubes of their mean distances from the Sun (Fig. 1.5). In Kepler's own description of this harmonic relation, written in 1618: "[It] agreed so perfectly with the data which my seventeen years of labor on Tycho's observations had yielded, that I thought at first I was dreaming."[28]

Invisible Powers Move the Planets

Kepler thought that the Sun governs the motion of the planets. In the first edition of his *Mysterium cosmographicum*, published in 1596, he had noticed that the more remote planets move more slowly, suggesting that either the individual soul that moves each planet is less active at greater distances from the Sun or that one moving soul in the central Sun drives the planets with

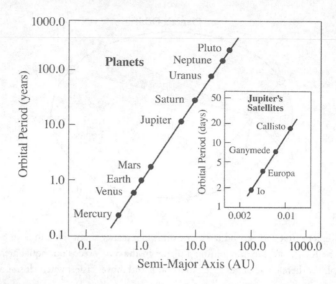

Fig. 1.5.   Kepler's third law The squares of the orbital periods of the planets increase with the cubes of their distances from the Sun. This relation is shown as a straight line in this logarithmic display; it also applies to Jupiter's four largest satellites shown in the inset.

less vigor the further the planet is. In the second edition of the book, published in 1621, Kepler substituted the mechanical word "force" for the magical term "soul." The Sun's force, which was responsible for moving the planets around their orbits, was supposed to diminish in strength with increasing distance as the intensity of light does, with the inverse square of the distance. The Sun was, after all, the source of the light that illuminates and warms the Earth.

What forces keep the planets moving in curved paths about the Sun? It might have something to do with magnetism, whose mysterious unseen force distributes iron filings about a bar magnet. In 1600, William Gilbert, physician to Queen Elizabeth I of England, authored a treatise with the grand title *De magnete, magneticisque corporibus, et de magno magnete tellure*, translated into English as *Concerning Magnetism, Magnetic Bodies, and the Great Magnet Earth*. In this work, Gilbert showed that the center of the Earth is itself a great magnet whose lines of force loop out to envelop the planet and explain the orientation of compass needles.

Drawing upon an analogy with the central magnetic source of the Earth, Kepler supposed that an invisible magnetic force emanates from the Sun, pushing the planets through space and controlling their motion. This indicated to him that: "The heavenly machine is a kind of clockwork, insofar as nearly all the manifold motions are caused by a most simple, magnetic, and material force ... given numerical and geometrical expression."[29] The further a planet is from the Sun, the weaker the solar magnetic force and the slower that planet's motion, as described by Kepler's harmonic relationship between the orbital period and distance from the Sun.

Kepler incorporated the Trinity of Father, Son, and Holy Spirit into his mystical interpretations of the Universe. The Sun, as God the Father, symbolized God's motive power emanating out to propel the planets. This invisible power extended throughout all of space, like the Holy Spirit. Supposing that the stars in the heavenly firmament represent Jesus Christ, he then proposed that the Sun, the stars, and the space between them are analogous to the Father, the Son, and the Holy Spirit.

His discoveries became the foundation for Isaac Newton's proposal that the invisible gravitational force of the Sun grasps the planets and holds them in place. He showed that the pull of gravity is universal, with an unlimited range and capacity to act on all matter, confining the Moon, planets, and comets in their trajectories. This meant that the same physical principles and mathematical laws describe motions everywhere, either up above in the Heavens or down here on Earth.

2. Gravity Guides Movement and Bends Space-Time

"Go Wondrous creature! Mount where science guides;
Go, measure earth, weigh air, and state the tides;
Instruct the planets in what orbs to run,
Correct old time and regulate the Sun
Go soar with Plato to the empyreal [heavenly] sphere."

Alexander Pope (1734)[1]

Newton's Unbroken Meditation

Isaac Newton was born on Christmas Day 1642,[2] shortly after the death of his father. Newton's mother Hannah remarried when he was three years old, and moved nearby to the home of her new husband, the wealthy and elderly clergyman, Barnabus Smith. Young Isaac stayed behind, in the house where he was born. His grandparents cared for him until his stepfather's death eight years later, when his mother returned together with three children of her second marriage.

At the age of 12, Isaac was sent away to the Free Grammar School in Grantham, where he learned Latin grammar and lodged with the local apothecary, who sold medicines and most likely gave Isaac an interest in chemistry. Later, at age 17, he returned to the family estate in Lincolnshire, and went on to Trinity College at the University of Cambridge in June 1661.

As a youth, Newton was exposed to Anglican (Church of England) and Presbyterian forms of worship, which vied for public interest and political control of England during Newton's early years. He was already a devout young man when entering college, and as an adult, he continued to hold an active, vigorous Christian faith and to embrace many of the local religious beliefs.

24

While a college student, Newton spent much of his time in religious activity, and he was not alone in this regard. All the Trinity students were required to read, critically examine, and know lengthy passages of the *Bible*, to attend daily sermons, and to participate in evening prayers. This was to be expected since about three quarters of the students then attending Trinity College were destined for a career in the Church of England.

Initially, Newton earned his keep at Trinity by serving the Fellows and wealthier students. He had to clean their boots, wait on tables, and empty their chamber pots. When these duties ended in 1664, Newton was elected a scholar at Trinity, but the next year the bubonic plague had spread to Cambridge and the University closed down.

Isaac therefore traveled to his ancestral farm and spent the next 18 months in intense meditation and thought. During this period he developed his theory of light and colors, began to think of gravity extending to the Moon, and realized how the forces that drive the planets must vary with distance from the Sun.

At the age of 24, Isaac came back to the University of Cambridge when it reopened in the spring of 1667, and that autumn was elected to a fellowship at Trinity College. Two years later he became Lucasian Professor of Mathematics, a position that he held for the next 32 years.

Newton just didn't like interacting with people, and was usually indifferent to them. He declined most invitations, avoided personal contact, and never traveled outside England. He was also reluctant to publish his findings, disliked controversy, did not desire "public esteem," and did not ask for help or invite collaboration in his investigations.

There is no evidence that the celebrated scientist ever loved a woman, and he apparently died a virgin at an age of 84. He was a lifelong bachelor who thought that sexual activity with women would corrode his spiritual purity, which had been suggested to him by a passage in the *Bible*.[3]

Newton was exceptionally curious, once sticking a long needle along the edge of his eye just to see how it worked and what would happen. On another occasion, he stared at the Sun so long that he could hardly see anything at all and was confined to his bed for days.

The great astronomer was capable of extraordinarily intense and prolonged mental concentration, and his incomparable discipline and rigorous work ethic continued throughout his life. He has attributed his significant accomplishments to this sustained thought and meditation, rather than moments of inspiration or genius.

Isaac was a detached, solitary loner, an isolated and self-contained intellect, a bit obsessed, notoriously absent-minded, famously distracted, and frequently depressed. Most of his life was spent as a reclusive Don at Cambridge, where he could work undisturbed, immersed in introspection, solitude, secrecy, and study.

Newton designed and built the earliest known reflecting telescope, which employed a mirror rather than a glass lens to gather and focus light (Fig. 2.1).[4] Isaac ground the metal mirror, constructed the tube, and used the completed telescope to observe the four large moons of Jupiter and the phases of Venus. When news of the novel instrument reached London in 1671, it earned him election as a Fellow of the Royal Society. At about the same time, Newton was developing his theory of light and colors, and communicating it to the Royal Society.[5] When he dispersed sunlight by passing it through a glass prism, Newton found that light is composed of different colors that were bent by the prism to different extents.

He then discovered the laws of motion and invented universal gravity, "the offspring of silence and unbroken meditation."[6] As he showed, gravity is always out there trying to pull us down. It keeps our feet on the spherical Earth, so we don't fall off it, and enables us to rotate with the spinning planet. The atmosphere and oceans are similarly held close to the Earth by its relentless gravitational pull. And it is gravity that explains how things fall. Every object on the Earth falls in just one direction, straight down toward the ground, and any undisturbed body will fall with uniform acceleration. The further it falls the faster it moves, at the same rate regardless of its mass.

Newton thought that the primordial stuff of the world was the result of God's existence, and that God deliberately created the ordered Cosmos by an act of will.[7] He also thought that the discovery of cosmic order reveals the mind of the Divine Creator, and that finding this order and discovering God's design was the best way to convince unbelievers of the existence of God. Newton did not, however, believe that the entire Universe was God, which would have been heretical at the time.

It has been said that Newton's religious views might not have had all that much impact on his efforts in astronomy, since he was able to compartmentalize his interests, and that he endorsed the separation of science and religion. These views are badly mistaken. In Newton's mind, there was

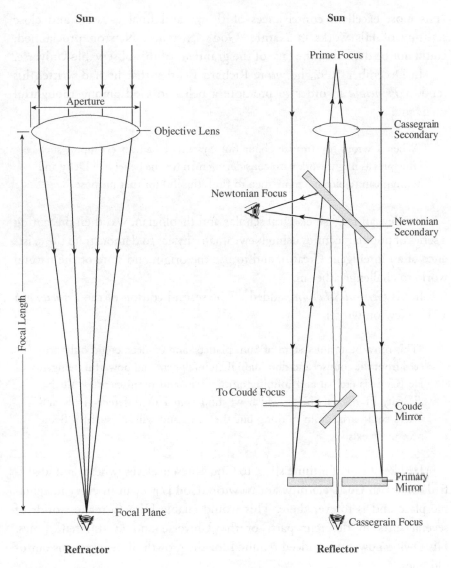

Figure 2.1 **Telescopes** Light waves that fall on the Earth from a distant object are gathered and focused to a point by the lens (*left*) or the mirror (*right*) of a telescope.

considerable overlap between God and investigations of the natural world, between theology and cosmology.

To Newton, the goal of natural philosophy was to understand God, the Divine Creator of the rationally ordered Universe, through knowledge of

"His most excellent contrivances of things and final causes," and close scrutiny of His works in Nature. God's existence, Newton proclaimed, could not be denied in the face of the grandeur of this observable Universe.

In December 1692, he wrote Richard Bentley that he had written his renown *Principia* in order to promote a belief in God among thoughtful men, and:

> "When I wrote my treatise about our System, I had an eye upon such Principles as might work with considering men for the belief of a Deity and nothing can rejoice me more then to find it useful for that purpose."[8]

Bentley, an English classical scholar and theologian, was then presenting a series of popular lectures using Newtonian physics to demonstrate the existence of an intelligent Creator, and to use the origin and state of the natural world to challenge atheism.

In his *General Scholium*, added to the second edition of the *Principia* in 1713, Newton wrote:

> "This most beautiful system of Sun, planets, and comets, could only proceed from the counsel and dominion if an intelligent and powerful Being... He [God] is eternal and infinite, omnipotent and omniscient... All that diversity of natural things which we find suited to different times and places could arise from nothing but the ideas and will of [such] a Being necessarily existing."[9]

He likened all of infinite space to God's presence everywhere, and identified time with God's eternity. For Newton, God is present in every imaginable place and is forever there. This would enable Him to make worlds of several sorts in different parts of the Universe and at different times. His Divine power was indeed required for the growth of new systems out of old ones.

Newton urged everyone to devote themselves to reading and study of the *Bible* He began an intensive, passionate study of theology as a young man around 1670, and devoted exceptional efforts to understand the Divine until the end of his life in 1727. Newton never doubted the existence of God, and had a robust faith that gave meaning to his life and work. He nevertheless questioned some practices of religion that he found contrary to his reason and to his exhaustive studies of ancient Scriptures and the early history of the Church.

His examination of the oldest extant Greek copies of *Holy Scripture* suggested that certain passages had been added around the 4th century after the birth of Christ. According to Newton, these false passages corrupted the true faith and misrepresented Jesus Christ. Although Christ was divine and worthy of worship, had died on the Cross and then been resurrected, he was not God himself, not equal to God. This meant to Newton that the doctrine of the Trinity, with the strict equality of the Father, Son and Holy Spirit, was not correct, and that he did not agree with Kepler's claim that the Cosmos has the Trinity embedded in it.

Newton's secrecy and habit of almost never publishing his extensive writings turned out to be a good thing, at least as far as his beliefs were concerned. If his denial of the Trinity had been published and became widely known, then Newton would have been immediately dismissed from the College of the Holy and Undivided Trinity at the University of Cambridge, and may have never written the *Principia* or acquired subsequent fame and position. His private beliefs were contrary to the faith of the Anglican Church, and would have also made Newton ineligible for his subsequent government positions at the Royal Mint in London.

At the time of Newton's election as the Lucasian Professor of Mathematics at the University of Cambridge, in 1669, the Fellows of Trinity College were required to become ordained priests in the Church of England within seven years. Although he was a sincere member of the Church, and publically participated in its religious services and practices throughout his life, Newton was not prepared to give up his academic freedom and take Holy Orders. Fortunately, the English king, Charles II, issued a royal decree in 1675, just when Newton's seven years was about to elapse, which exempted the Lucasian Professor from needing to take Holy Orders.

Newton wanted to purge Christianity of irrationalities, which he thought had been introduced by the Roman Catholic Church, and he disagreed with many Catholic practices, as many did in the English society of the time. Like Martin Luther and other participants in the Protestant Reformation, Newton thought that Catholics had perverted the true faith by the use of indulgences, compulsory confession, and the introduction of idolatrous beliefs and practices that appeal to superstition and mystery. To Newton, this false idolatry included the magical use of the sign of the cross, the cult of the Virgin, the worship of Christ as God, the veneration of saints and their relics, and the supremacy of the Pope.

Perhaps because of his unprecedented scientific discoveries, Newton has often been portrayed as a man of incessant reason; a person solely dedicated to science or natural philosophy as it was then called. This is far from the truth. He devoted significant time and effort to other subjects including alchemy, theology, the early Christian Church, and interpretations of *Biblical* prophecy. Newton also spent much of his life trying to understand the origin of the elements and the eternal mysteries of health and human death by examining what he considered to be mystical clues left by God.

The renowned economist John Maynard Keynes, who purchased most of Newton's alchemical papers, wrote that Newton's deepest instincts were "occult and esoteric," and that he was "the last of the magicians" who "looked on the whole Universe and all that is in it *as a riddle*, as a secret. By concentration of mind, the riddle, he believed, would be revealed to the initiate."[10]

In his lifetime, Newton's contemporaries viewed his scientific achievements with awe and admiration, and he became the most famous of men, celebrated as a great and rare genius. Upon his death in 1727 he was buried with ceremonious pageant and great pomp in Westminster Abbey, London, when two dukes, three earls and the Lord Chancellor carried the coffin. His elaborate tomb, erected in 1731, includes symbols of his discoveries — a prism, a reflecting telescope, and the Sun, planets and comets.

The Principles

When the English astronomer Edmond Halley visited Newton at the University of Cambridge in August 1684, Halley asked him about the unsolved problems of planetary movements around the Sun. Newton replied that he had already found a solution but mislaid it.[11]

Newton had confirmed Kepler's conclusion that a planet moves in an elliptical orbit under the influence of a force originating at a focus of the ellipse, and that the force must decrease in strength as the inverse square of the distance from that focus. Since the Sun was at the focus of the elliptical planetary orbits, it must exert a gravitational force on the planets that varies with distance from the Sun in that way. Such a decrease of the Earth's gravitational force with distance from the Earth also explains the Moon's orbital motion.

Newton was also probably aware of Kepler's *Astronomiae Pars Optica (The Optical Part of Astronomy)*, published in 1604, in which Kepler reasoned that the intensity of light decreases with the inverse square of the distance. As either gravity or light move away from their source and fill the increasing volume of space, they get weaker in the same way and exhibit a similar decrease in either force or intensity.

Encouraged by Halley, Newton spent 18 months of intense labor to redo his work and write his great treatise, the *Philosophiae naturalis principia mathematica*, or the *Mathematical Principles of Natural Philosophy*, commonly known as the *Principia*. It was presented to the Royal Society, which withdrew from publishing it owing to insufficient funds, so in 1687 Halley saw the book through the press and was paid for his work by the Royal Society in the form of 50 copies of Francis Willughby's *De Historia piscium* (*The History of Fishes*) instead of 50 pounds.

The first *Book I* of the *Principia* concerned *The Motion of Bodies*, where Newton described space and time and introduced the concepts of mass, force, and quantities of motion, including parameters known today as *centrifugal force*, *momentum* and *inertia*. As the title *Principia*, or *Principles*, suggests, a few simple natural laws could be used to describe any moving object, from a tossed stone to the Moon, comets, and planets. If the object was at rest, it stayed at rest, and if it was moving in empty space, it continued in motion at the same speed and in the same straight direction, unless it was compelled to change that state by external forces impressed on it. The new motion would be proportional to the motive force impressed, and would be in the direction of the force.

In *Book II* Newton examined the effects of a *Resisting Medium* on a body's motion, such as air, water, or friction.

The last *Book III* entitled *De mundi systemate* (*On the System of the World*) unified the Earth and Heavenly bodies through the principle of universal gravitation. It discussed the consequences of gravity to known worlds, especially for astronomical observations, and described how the unseen powers of gravitation thread their way across space, guiding material objects along invisible but determined paths. When combined with the rules of motion, universal gravitation describes the movement of cosmic objects everywhere in the Universe.

Since the Earth's gravitational force is present even at the top of the highest mountains, Newton imagined that it extends all the way to the Moon. He demonstrated that this force, diminished by distance to the Moon, pulls it into ceaseless motion around the Earth.

This is the universal force of gravity that operates between all bodies and pulls any two material objects together. The attraction is centered in each object, increases with their mass, and strengthens with proximity, as the inverse square of the distance. [Mass is an intrinsic aspect of an object, different from its weight that alters with distance from the main source of gravity, but back in Newton's time, mass and weight were assumed to be equal.]

How did Newton arrive at the concept of universal gravitation? He emphasized that his laws were discovered by observations and experiment, and not by rational deduction, uncertain conjectures, or complex mathematics. It was enough that these laws allowed successful predictions to be made for observed phenomena. In the second edition of his *Principia*, published in 1713, he therefore famously wrote:

> "I have not as yet been able to deduce from phenomena the reason for these properties of gravity, and I do not feign hypotheses [hypotheses non fingo]… And it is enough that gravity really exists and acts according to the laws that we have set forth and is sufficient to explain all the motions of the Heavenly bodies and of our sea."[12]

In other words, Newton explained *how* material bodies move; he subsequently attributed the *why* they move to God.

In old age, he also told a few friends of watching an apple falling in his orchard, which reminded him of the power of gravity whose pull influences the motion of all falling bodies. Newton noted that the Moon resembles a projectile fired from the Earth and constantly falling toward it, but moving so fast that it always keeps the same mean distance from the Earth. If the Moon moved any faster, it would travel out into space, never returning to the Earth, and if the Moon moved any slower, the Earth's gravity would pull it to the ground.

Like our Moon, the planets travel along curved trajectories, and in this case Newton proposed that it is the Sun's unseen gravitational force that guides the planets in their motion. If left to themselves, the planets would not move in a closed path, and instead travel along straight-line paths. But they are not "left to themselves" because an external force is exerted on them by the Sun. The relentless pull of the Sun's gravity holds the planets in their rounded paths, keeping them balanced and suspended rather than falling into the Sun or moving off into interstellar space.

The motions of comets soon provided a demonstration of this universality.

Return of the Comet

Unlike the planets, the comets can appear almost anywhere in the sky, remain visible for a few weeks or months, and then vanish into darkness. For centuries, no one knew where comets came from, where they went, or when they might be expected to appear.

The unexpected arrival of the awe-inspiring comets seemed to upset the natural order of the eternal, unchanging Heavens. Their appearance was thought to foretell wars and other disasters on Earth, such as the assassination of Julius Caesar, the Norman conquest of England, and the Turkish conquest of Constantinople. As Shakespeare declared:

> "When beggars die there are no comets seen;
> The Heavens themselves blaze forth the death of princes."[13]

The mystique, fear, and superstition associated with comets were largely removed by Edmond Halley in his *Astronomiae Cometicae Synopsis* of 1705.[14] He compiled a large number of previously recorded observations of comets from old and rare sources, and used them to calculate each comet's orbit using Newton's theory of gravity. Halley found that twenty-four bright comets seen in previous centuries moved around the Sun in very elongated, elliptical orbits as Newton had speculated.

Halley went further, and predicted that at least one bright comet had been seen more than once during several trips around the Sun and that it would be seen again. He had found that the observed trajectories of the comets seen in 1456, 1531, 1607 (noted by Kepler and Shakespeare), and 1682 were very similar, and concluded that they were four apparitions of the same comet, which appeared at 76-year intervals when coming close to the Sun. Halley predicted that the bright comet would return in 1758, and if that happened he hoped that it would be remembered that an Englishman had predicted its reappearance. He was criticized for placing the date of the comet's return so far in the future that he would most likely not be alive to see it. [Halley died in 1742 at the age of 85.] But the comet came as predicted, and was re-discovered on Christmas night in the expected year, the first comet to arrive on schedule and now the most celebrated of the comets, bearing Halley's name (Figs. 2.2, 2.3).

The earliest apparition of Halley's comet, established with confidence from Chinese chronicles, dates back to 240 BC. Since then, all thirty-two of its passages near the Sun have been retraced in the ancient or modern records of astronomers. Halley's comet has now moved away from the Sun into cold icy darkness where it cannot be seen, but it is heading back toward us along an elongated orbit that will bring it into sight in 2061. The Sun's heat will then vaporize the outer parts of the small icy world to enlarge it and perhaps form an extended tail that could be as long as the distance between the Earth and the Sun.

Figure 2.2   Comet Halley in 1759 AD This Korean record of comet Halley was made during the comet's first predicted return in 1759 AD. (Courtesy of Il-Seong Na, Yonsei University, Seoul.)

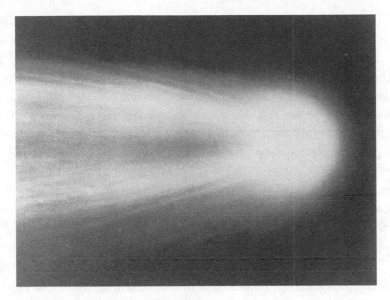

Figure 2.3   Comet Halley in 1910 The head region or coma of comet Halley observed on 8 May 1910 with the 1.5-meter (60-inch) telescope on Mount Wilson, California. The comet's tail flows to the left, away from the Sun. (Courtesy of the Hale Observatories.)

The development of telescopes in the 19th and 20th centuries resulted in the discovery of many comets that cannot be seen with the unaided eye. When their orbits were determined it was found that they are all tiny invisible denizens of the outer Solar System, residing far away but still within the Sun's gravitational embrace. They spend most of their unseen life in icy hibernation far from the Sun, and only become visible when a passing star tosses them into the realm of the inner planets, where the Sun's heat vaporizes their surface ice and they grow large enough to reflect visible amounts of sunlight. Up to 200 billion of them are thought to move about the Sun in the remote unseen darkness, some of them halfway to the nearest star.[15]

Newton the astronomer thought that God's intervention was needed to prevent the Sun and stars from being pulled together by gravity, and that the observed laws of Nature, which incidentally proved the existence of such an all-powerful God, did not prohibit such "miracles". To Newton, all motion would eventually run down and stop, and everything in the currently observable material world would inevitably wear out in time and cease to exist. But the free and powerful "Lord God," who has a "propensity for action," could correct the situation at all times anywhere in the observable Universe.

This active God, this Divine Will, was required to sustain the Universe and to assure the renewal of systems in it. So the Universe was not only created by the will of God; it is also preserved, governed, maintained and continued by God's active intervention, according to his will and wishes.

Throughout much of his adult life, Newton also provided interpretations of *Biblical* prophecies that he found in ancient Greek versions of *The Revelation of St. John the Divine*. He thought that God's dominion over a fallen people explained the plague and barbaric invasions of Europe, and that things did not look good for Catholics or believers in the strict Trinity at the future Second Coming, which would bring Christ's judgment over all the kingdoms of the Earth.

A Universal Truth

It was the prediction of the existence of Neptune that convinced astronomers, and just about anyone else, of the far-reaching power of universal gravitation. Another remote planet, Uranus, had been accidentally discovered in 1781; by William Herschel who initially thought it was a comet.[16] Uranus had been detected by professional astronomers and mistaken for a star on no less than 22 occasions during the century that preceded the realization that it was a planet. These additional observations were combined with the post-discovery ones to determine Uranus' trajectory and calculate its future position. Before long it was found that the planet was wandering from its predicted path.

A large, unknown world, located far beyond Uranus, was evidently producing a gravitational tug on the planet, causing it to deviate from the expected location. Two astronomer-mathematicians, John Couch Adams in England and Urbain Jean Joseph Le Verrier in France, independently specified the location of the planet by using Newton's theory of gravitation to explain the observed motions of Uranus.[17]

Adams, a recent graduate from the University of Cambridge, finished his work first, deriving a precise position of the planet in mid-1845. He left a summary of his results with the then Astronomer Royal, George Biddell Airy, who did not feel compelled to look for the unknown world. Le Verrier finished his best calculations about a year later, and, unlike Adams, published his results. The two astronomers had arrived at nearly identical locations for the unseen planet.

When Le Verrier's memoir reached Airy, he persuaded James Challis, Professor of Astronomy at the University of Cambridge, to make a search for the undiscovered planet. For a variety of reasons, Challis began the investigation slowly, and Le Verrier had in the meantime sent his results to the Berlin Observatory where Johann Gottfried Galle and his student Heinrich Louis d'Arrest found the planet.[18] They identified it on the first night of their search, on September 23, 1846, using a 0.23-meter (9-inch) refractor; it was located within a degree of both Adams' and Le Verrier's predicted positions. Only later did Challis realize that he had previously observed the planet twice when beginning his own search.

The discovery of the new planet, named Neptune after the Roman god of the sea, was acclaimed as the ultimate triumph of Newtonian science. It resulted from mathematical calculations, based on Newton's theories, of the effects of a previously unknown planet whose gravity was pulling Uranus from its predicted place. If proof were needed, this achievement certified the validity of gravitational theory.

This verification provided the foundation and impetus for all of science to come. Universal scientific laws are thought to apply anywhere throughout the observable Cosmos. Given present circumstances, these laws can be used to predict what will happen in the future. For astronomers, these universal and eternal physical truths are always present, even if they are as yet undiscovered.

Everything that an astronomer observes must obey the natural laws that apply to its particular situation. No exceptions are permitted. You cannot break the law in astronomy, where the truth prevails and no one can lie, at least for very long. All new discoveries, and every explanation of them, must be verifiable, and they are always subject to question and doubt.

In order to obtain objective, repeatable, and verifiable physical truths, astronomers cordon off and scrutinize a tiny portion of the vast and largely unknown Cosmos. This closed-off part is assumed to be isolated from all external influences, and every prediction about them is hedged with the proviso "other things being equal" or "no unexpected external forces allowed." As a result, the astronomers' truths are always qualified by boundary conditions, and their knowledge of the Universe is forever limited and never complete.

Towards the end of his life, Isaac Newton humbly recognized these limits to current knowledge, exclaiming: "I have been like a boy playing on the seashore, and diverting myself in now and then finding a smoother pebble or a prettier shell than ordinary, whilst the great ocean of truth lay all undiscovered before me."[19]

In the memorable words of the 20th century Chilean poet Pablo Neruda: "In this net it's not just the strings that count but also the air that escapes through the meshes."[20] The astronomers' nets are always open to unknown possibilities that might be revealed by future observations. Their lives are often driven by a continual quest for these unseen, mysterious things, and they are always carrying us beyond the existing boundaries of current astronomical vision, sometimes beyond the limits of known natural laws.

The unexpected behavior of the planet Mercury provides an early example of this ongoing quest. Instead of returning to its starting point to form a closed ellipse in one orbital period, Mercury moves slightly ahead in a winding path that was first described by LeVerrier in 1859, when he attempted to use Newtonian gravitation to account for observations of the planet's transits in front of the Sun. LeVerrier subsequently attributed the mysterious motion to the gravitational pull of an unknown planet orbiting the Sun inside Mercury's orbit and moving ahead of it. The hypothetical planet named *Vulcan* was never reliably detected.

As we shall next see, the cause of Mercury's unexplained motion remained a mystery until 1915 when Albert Einstein described how gravity works. According to his explanation, massive objects produce a curvature in nearby space-time, and that bending, twisting and distortion is gravity. The prediction of consequences of this curvature made him famous and still does.

Einstein's Cosmic Religious Feeling

Albert Einstein was born on March 14, 1879 at Ulm, in southern Germany, but after the age of four grew up in Munich. As a child of four or five, Einstein's father showed him a compass needle that "did not fit into the usual explanation of how the world works," leading him to conclude that "there must be something deeply hidden behind everything."[21] A mysterious and unseen agent was making the compass needle move without touching it, and that unknown something inspired awe and wonder in Einstein.

Albert also had a deep, lifelong love of music. He played the violin, and his favorite composers included Bach, Mozart, and Vivaldi. Einstein often thought in music, daydreamed in it, and saw his life in terms of music.[22]

The Einstein family moved to Milan, Italy in 1894, but Albert stayed in Munich to finish high school, and then went to Switzerland to continue his education. After failing his college entrance exams at the Zurich Federal Polytechnic School, he went to a high school in Aarau, the German speaking part of Switzerland, which enabled him to enroll in 1896 in a four-year course at the Zurich Polytechnic that was designed to produce high school teachers. [In 1911 the school was given its current name Eidgenössische Technische Hochschule, or ETH for short.]

At Zurich, Einstein began life-long friendships with Michele Besso and Marcel Grossman, and also became romantically entangled with Mileva Maric, a fellow student at the Polytechnic. In 1901, they had a child out of wedlock, a daughter, whose real name and fate remain unknown. She was probably put up for adoption, but Einstein never saw her. Albert married Mileva in a civil ceremony in 1903 without the presence of a rabbi or priest, and to the discontent of both sets of parents. [His family was Jewish, and she was a member of the Greek Orthodox Church.] Two sons were born to the couple, Hans in 1904 and Eduard in 1910.

In 1900 Einstein graduated from the Zurich Polytechnic with a diploma to teach mathematics and physics in high school. He applied for a job as a university lecturer and was turned down, and then as a high school teacher and was rejected again. After two years of failure to find a job, Marcel Grossman's father helped Einstein obtain a job in Bern at the Swiss patent office, where he stayed for the next seven years.

Einstein evaluated patent applications for electromagnetic devices, which he enjoyed, and it left him time to write scientific papers, which he contributed to the *Annalen der Physik*. These papers were written without any significant contact with the physics community, and they attracted little notice amongst them. After all, an unknown clerk in a Swiss patent office wrote them.

But then, in 1905, Einstein published a wide-ranging series of original ideas that could not be ignored, and after that his professional career began to rise. He moved from university teaching as a Privatdozent in Bern in 1908, to Associate Professor at the University of Zurich in 1909, to Full Professor at the University of Prague in 1911, and on to Full Professor at the ETH in 1912.

In 1913, Einstein was offered a Full Professorship without teaching obligations at the University of Berlin, under the aegis of the Prussian Academy of Science, which he held for the next eighteen years. He apparently had enough of teaching, and also his unhappy marriage. Mileva and their two sons remained in Zurich, and after five years apart, the couple divorced.

In late 1915, Einstein completed his *General Theory of Relativity*, which explained an unexpected aspect of Mercury's motion as the result of the Sun's curvature of nearby space, and predicted that this curvature would cause the bending of starlight passing near the Sun. After the effect was observed during a solar eclipse in 1919, Einstein became a legend, almost overnight.

Einstein brought an "other worldly" order to the Universe. He was a new Moses come down from the mountain to bring the law and a new Joshua controlling the motion of heavenly bodies. He spoke in strange tongues, and the stars demonstrated the truth of his sacred message.[23]

For the rest of his life, Einstein basked in international fame; abruptly ended his creative scientific work; and became more noted for his support of social justice, his defense of the weak and oppressed, and important political activities.

As an example, when Adolf Hitler came to power, Einstein was in the United States, where he settled down as a Professor at the Institute for Advanced Study in Princeton, New Jersey, while also becoming a citizen of the United States in 1940. The preceding year Einstein sent a letter to the United States President Franklin D. Roosevelt alerting him to the possibility of setting up a nuclear chain reaction by which "extremely powerful bombs of a new type may thus be constructed."[24] He suggested that Hitler's forces might be developing such weapons, and recommended that the United States begin making one. This helped lead to the Manhattan Project in which the best scientific minds in the country moved to Los Alamos, New Mexico and created the first atomic bomb.

Einstein did not lead a lonely life, thanks to the loving care of his cousin Elsa Löwenthal. They married in 1919, and parted without children on Elsa's death in 1936. But Einstein never did get along well with women. In 1955, shortly after the death of his friend Michele Besso, he wrote to the Besso family, stating: "What I most admired in him as a human being is the fact that he managed to live for many years not only in peace but also in lasting harmony with a woman — an undertaking in which I twice failed rather disgracefully."[25]

He instead spent his entire life in understanding and explaining how the natural world works. As Einstein expressed it: "Out yonder there was this huge world, which exists independently of us human beings and which stands before us like a great, eternal riddle, at least partially accessible to our inspection and thinking. The contemplation of this world beckoned like a liberation."[26]

Einstein had an unshakeable conviction that there is a hidden order and unity in nature, which the probing mind can partially comprehend. As he expressed it, we have "a knowledge of the existence of something we cannot penetrate, [manifesting itself in] our perceptions of the profoundest reason and the most radiant beauty, which only in their most primitive forms are accessible to our minds — it is this knowledge and this emotion that constitute true religiosity; in this sense, and in this alone, I am a deeply religious man."[27]

Einstein did not believe in the personal God who cares for human beings and directly interacts with them, rewarding the righteous, punishing the wicked, and providing comfort after death. But he considered himself a very devout man, with a deep and profound belief in "his God, his Lord." To him, the Divine is not isolated from the physical world, but instead revealed by it.

Throughout his life, Einstein was enlivened by this "cosmic religious feeling" that included an awe, admiration, and wonder for the Universe itself, for its subtlety, beauty, and elegance. In his own words:

> "The individual feels the futility of human desires and aims, and the sublimity and marvelous order which reveal themselves both in Nature and in the world of thought. Individual existence impresses him as a sort of prison, and he wants to experience the Universe as a single significant whole. ... In my view, it is the most important function of art and science to awaken this feeling and keep It allve In those who are receptive to it."[28]

And on another occasion: "The fairest thing we can experience is the mysterious. It is the fundamental emotion that stands at the cradle of true art and true science. He who knows it not and can no longer wonder, no longer feel amazement, is as good as dead, a snuffed out candle."[29]

Throughout his life, Einstein was a man apart, a lone traveler. As he confessed in 1930: "I have never belonged to my country, my home, my friends, or even my immediate family, with my whole heart. ... I have never lost a sense of distance and a need for solitude."[30] This independence of mind, this apartness, led to that amazing sequence of papers in 1905 that jump-started his career and transformed our understanding of light, space, time, mass and energy.[31] He then moved on to gravity.

Einstein Discovers how Gravity Works

We can't see the force of gravity, and Newton didn't know how it was exerted. Its unseen hand was supposed to operate everywhere across the space between all bodies that do not touch each other. Newtonian gravity therefore resembled a mystic, occult quality, a ubiquitous presence.

Einstein amended Newton's laws to explain how gravity works. He supposed that a massive body like a star bends nearby space-time into the curvature of an embrace, giving it a shape and form. Gravity works by this bending, twisting and distortion of space-time. In effect, mass tells nearby space-time how to curve and curved space-time tells neighboring matter how to move. But such curvature effects are only noticeable in extreme conditions near a very massive, cosmic object like a star, and the differences between Newton's and Einstein's theories of gravity are indistinguishable in ordinary circumstances on the Earth.

The new theory arose because of an exceedingly tiny, unexplained aspect of the planet Mercury's motion around the incredibly massive Sun. Instead of returning to its starting point in one orbital period, Mercury moves slightly along a path that can be described as a rotating ellipse (Fig. 2.4). As

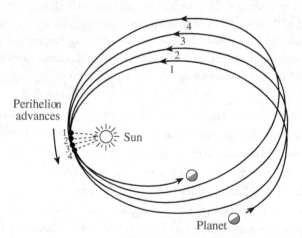

Figure 2.4 Mercury's motion Instead of always tracing out the same ellipse, the orbit of Mercury pivots around the focus occupied by the Sun. Albert Einstein explained this movement by extending Newton's theory of gravity to include the Sun's curvature of nearby space-time.

a result, the point of Mercury's closest approach to the Sun, the perihelion, advances by a small amount, just 43 seconds of arc per century, beyond that which can be accounted for by planetary perturbations using Newton's theory of gravitation.

Although discovered by the French mathematician Urbain Jean Joseph Leverrier in 1859,[32] the unexplained motion of Mercury remained a mystery for more than half a century, until 1915 when Einstein identified its cause in his *General Theory of Relativity*.[33] He showed that Mercury is directed along a path in curved space-time. It is something like watching the slow arc of a bird gliding on unseen winds.

When Einstein found that his calculations agreed with the unexplained observations of Mercury, it changed his life. He wrote to his friend Michele Besso that: "My wildest dreams have been fulfilled," and subsequently confessed that "for a few days, I was beside myself with joyous excitement," and that it "had given him palpitations of the heart, with a feeling that something actually snapped in him."[34]

Like any good scientist, Einstein realized that his *General Theory of Relativity* had to be verified by definitive tests of other consequences, and in the very paper that explained Mercury's unexpected motion he predicted that the curvature of space-time would also deflect starlight from a straight-line path.[35] Newton had previously speculated that massive bodies might bend nearby light rays if light has mass; but when Einstein took curvature into account the expected deflection was doubled, to 1.75 seconds of arc for a light ray grazing the Sun's edge.

Einstein's novel concepts of 1915 were made during World War I, that began on July 28, 1914 and lasted until the armistice of November 11, 1918. The war pitted Great Britain and its allies against Germany, and as the result of the warfare the English people suffered extraordinary hardship and loss of life. There were wide-spread anti-German feelings in England, and communications with enemy scientists were forbidden. So, it is quite amazing that the *General Theory*, which was conceived and published by a German scientist, became known to English astronomers who arranged to test it during the darkest days of the war.

One of the principal actors in this drama was the Quaker astronomer A. S. (Arthur Stanley) Eddington, who opposed any war and advocated international cooperation amongst astronomers. He therefore wrote in 1916 that "the pursuit of truth, whether in the minute structure of the atom or the vast system of the stars, is a bond transcending human differences — to use it as a barrier fortifying national feuds is a degradation of the fair name of science."[36]

It was Willem de Sitter, in the neutral Netherlands, who sent Eddington articles in English about the new amendments to Newton's gravitation theory, which were then published in England's *Monthly Notices of the Royal Astronomical Society*.[37] As the result of Eddington's informal expositions and a major presentation given in his *Report on the Relativity Theory of Gravitation* (London: Fleetway Press 1918) many English astronomers learned of the new concept. And already in 1917, the Astronomer Royal, Frank Dyson, was actively seeking support for an expedition to test Einstein's light-bending prediction during a total solar eclipse on May 29, 1919.[38]

If Einstein was right, the curvature of space-time would bend the light of stars passing nearly behind the Sun and spread their apparent positions apart like a gigantic magnifying lens (Fig. 2.5). The effect could be established by comparing the observed positions of adjacent stars seen near the

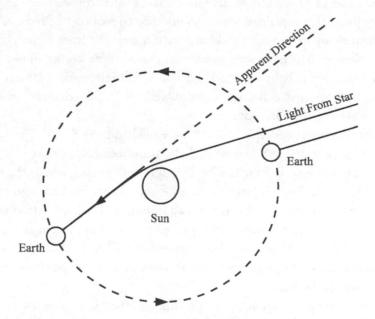

Figure 2.5 The Sun bends starlight The massive Sun creates a curvature in nearby space-time, which bends the path of starlight passing near it.

Sun during a solar eclipse with those seen long before or after and far from the Sun's location in the sky. When near the Sun the observed separations between the stars are greater than when the stars are far away from the Sun.

In June of 1918 there was the possibility that Eddington might be sent to jail due to his refusal to participate in the Great War. That's when the Scottish astronomer Frank Dyson came to the rescue of his old friend and former colleague at the Royal Greenwich Observatory. Dyson, with connections to the Admiralty as Astronomer Royal, testified before Eddington's Appeals Tribunal, stating that Eddington was uniquely qualified to carry out distant, important eclipse observations the next year. On hearing that such an eclipse might not be seen again for a very long time, the Tribunal declared that Eddington's work continued to be of national importance and provided him a 12-month exemption on the condition that he participate in the eclipse expedition, which he did. By that time, the war was over, and the issue of Eddington's military service vanished without a review of his concientious objection.

Interest ran high in Britain, and funding was obtained for a joint Greenwich Observatory — Royal Society expedition to measure the deflection of starlight during the total solar eclipse of May 29, 1919, when stars could be seen near the darkened Sun.

English expeditions were sent to observe it from the small Principe Island off the coast of West Africa and from Sobral, Brazil. In the last evening before sailing to Principe, E. T. Cottingham, who was to accompany Eddington, asked Dyson what would happen if they found *twice* Einstein's predicted deflection. Sir Frank replied: "Then Eddington will go mad and you will have to come home alone."[39]

Clouds and rainy weather interferred with the eclipse observations from Principe, and telescope difficulties compromised the Sobral ones. Measurements were made of the separation of stellar positions on photographs taken during the eclipse from both locations, and compared to those obtained long before when the same region of the sky was nowhere near the Sun. Deflections were observed, and they favored Einstein's explanation of how gravity works, but the imprecise measurements did not confirm it with complete definitive confidence.

In his report given at a joint meeting of the Royal Society and the Royal Astronomical Society,[40] on November 6, 1919, Dyson nevertheless declared that there was no doubt that they verified Einstein's law of gravitation, and it turned out that his confidence was ultimately justified. The Sun's curvature of nearby space-time has now been measured with increasingly greater precision for nearly a century, confirming Einstein's prediction to two parts in a hundred thousand, or to the fifth decimal place.

The apparent confirmation of the bending of starlight by the Sun in 1919 brought Einstein international recognition and capitulated him into the public limelight. Members of the press were present when Dyson announced the eclipse results, and the headlines of the London *Times* on the next day read "Revolution in Science — New Theory of the Universe — Newtonian Ideas Overthrown." Half way down the page was a second heading: "Space Warped." Two days later, the *New York Times* called Einstein's theory one of the greatest, perhaps the greatest, achievement in the history of human kind.

Einstein resembled ancient prophets and saints who could understand what others could not, and tell them about it. He gave them something to believe in, and these new things were not the way they used to be. Time was no longer exact; space was curved and warped; light had weight; and stars were not where they were supposed to be. Invisible gravitational waves might even be rippling unseen throughout space.

Gravity Makes Waves

Astronomers had to wait a very long time before the confirmation of Einstein's 1916 prediction of gravitational radiation. The vibrations are so weak and their interaction with matter so feeble that Einstein questioned whether they would ever be detected. As it turned out, orbiting neutron stars and merging black holes can produce detectable gravity waves. These ripples carry away energy from their source, and stretch and compress the space they pass through.

The existence of gravity waves wasn't demonstrated until 1974, when the two American radio astronomers Russell A. Hulse and Joseph H. Taylor, Jr. decided to replace the strip-chart method of discovering radio pulsars with digital computer techniques that provided the signal processing needed for a sensitive pulsar search. Their extensive survey using the 305-meter (1000-foot) Arecibo radio telescope in Puerto Rico resulted in the discovery of the first known pulsar with a companion.[41]

The pulsar's period of just 0.059 seconds rhythmically increases and decreases by very small amounts every 7.75 hours, as the result of its orbital motion about another neutron star that does not emit detectable radio pulses. As Hulse and Taylor predicted, precise timing of the radio pulses permits measurements of the relativistic orbital parameters and the masses of the pulsar and its silent companion. They weighed in at 1.44 and 1.39 times the mass of the Sun, as would be expected for two neutron stars.

After routine computer analysis of about 5 million pulses during a four-year period, Taylor and his colleagues found that the orbital period was slowly becoming smaller. The two stars were drawing closer and closer together, approaching each other at the rate of about 1 meter per year. This is the change expected if the orbital energy of the two neutron stars is being carried away by gravitational waves.[42] It demonstrated the existence of gravitational waves and opened up new possibilities for the study of gravitation — for which Hulse and Taylor received the 1993 Nobel Prize in Physics.

Direct detections of gravitational waves, with measurements of their waveforms, were announced on February 11, 2016 and June 15, 2016,[43] about a century after Einstein's prediction. Two identical instruments known as an Advanced Laser Interferometer Gravitational-Wave Observatory, abbreviated LIGO, were used to sense the distortions that occur when gravitational waves pass through these detectors and the underlying ground. One LIGO instrument is located in Washington State and the other in Louisiana, and gravity waves have to be observed in each of them to confirm a single detection. Each detector has the shape of a giant L with two perpendicular legs that are both four kilometers long. Laser light is sent back and forth through the two legs by multiple reflections from mirrors located at the ends of each leg, but canceling each other out at the place where they meet.

When a gravitational wave passes through the observatory, there is a change in the relative length of the two legs, and the laser beams are no longer synchronized on arrival. An oscillation is then detected that resembles the chirp of a bird with a rapidly increasing pitch or "ring-down," from 35 to 250 cycles per second.

The two chirp-like signals have each been attributed to the coming together of two black holes that have merged into a single, larger black hole. In the first detection, one black hole was about 36 times the mass of the Sun, and the other about 29 solar masses; in the second detection, the merging black holes were about 14.2 and 7.5 solar masses. When they got near enough to each other, each pair spiraled together into a black hole, which weighed about 62 solar masses in the first instance and 20.8 times the mass of the Sun in the second one.

The change in the length of the legs was exceedingly small, by much less than the diameter of a proton, but the result is a very big deal. More than a thousand scientists are now working on the $1-billion LIGO experiment. Some people have dedicated their entire working life to it, from its construction to first run in 2002 and eventual success 14 years later. And despite the tiny observed signal, the black-hole merger that caused it temporarily radiated an enormous energy in the form of gravitational waves, more than all the stars in the observable Universe emitted as light in the same time.

You might say that it is about time that gravitational waves were detected. Einstein predicted their existence almost exactly 100 years before the LIGO result. Joe Weber was trying to detect them more than half a century ago, and in 1971 Stephen W. Hawking, the theoretical physicist and cosmologist at the University of Cambridge predicted they would be generated by colliding black holes.[44] Subsequent experiments using massive detectors similar to Weber's indicated his reported detection was a false alarm. As recently as 2014, reputed observations of gravitational-wave signatures in the cosmic microwave background, by BICEP2, were also discounted. But this time around, there is not one but two LIGO detectors that are widely separated on the Earth, and the fact that they both "heard" the same faint chirps gives added confidence in the result.

Rainer Weiss, Barry Barish, and Kip Thorne received the 2017 Nobel Prize in Physics "for decisive contributions to the LIGO detector and the observation of gravitational waves."

The European Gravitational Observatory has constructed an interferometer gravitational-wave detector, named Virgo, near Pisa, Italy; it began joint observations with LIGO in 2017. Both the two LIGO and the Virgo instruments have directly detected gravitational waves on several occasions; the radiation has been attributed to either the collision of a pair of neutron stars or to a binary black hole merger. Future experiments may provide additional insight to massive, binary black holes, confirm that gravitational waves travel at the speed of light, and even measure the rate of expansion of the Universe.

Walking in the Midst of Wonder

The man who predicted that gravity bends the path of light, and that unseen gravitational waves ripple through space, was driven by his belief in the rational, intelligible order of Nature, which we only partly see and must be humbled by. Whenever this motivation is absent, he thought, science degenerates and becomes uninspired.

Einstein also supposed that there is much more to human life than scientific knowledge, a rejuvenating dimension implied in his letter to Queen Elizabeth of Belgium. It includes: "As always, the springtime Sun brings forth new life, and we may rejoice because of this new life and contribute to its unfolding; and Mozart remains as beautiful and tender as he always was and always will be. There is, after all, something eternal that lies beyond reach of the hand of fate and all human delusions."[45]

Astronomers have similarly tapped into something eternal, which normally operates beyond the range of known perception. We all walk in the midst of this wonder.

3. Motion within Matter

"To see a world in a grain of sand,
And a heaven in a wild flower,
Hold infinity in the palm of your hand,
And eternity in an hour."

William Blake (1803)[1]

Everything in the material Universe is composed of countless unseen, moving particles. Some of them are atoms or molecules; others are sub-atomic particles like electrons, neutrons, or protons. Invisible atoms are moving about in the solid chair you are sitting in; unseen electrons travel along wires to light your room; and the molecules in the air you breathe are darting about with a speed that increases with their temperature. Large numbers of imperceptible, electrified particles are even moving out from the Sun, many of them enveloping the Earth.

Invisible, Immortal Atoms

What are things made out of? To find out, you might try breaking any object into smaller and smaller pieces, until you reach a stage when the smallest piece cannot be broken apart. That last step in this imaginary decomposition arrives at the *atoma*, a Greek word that means "unable to cut."

The belief that all visible objects are composed of tiny unseen, immortal atoms dates back thousands of years, to ancient Greek and Roman philosophers, and it has never been discounted. The incredible atoms are supposed to be in continual motion within all that exists throughout all eternity, coming together to create visible things and dispersing when objects fall apart. They are the former ingredients of all that existed in the past and will be the seeds of everything that might exist in the future.

As the great English astronomer Isaac Newton wrote in 1704:

"It seems probable to me that God in the Beginning formed Matter in solid, massy, hard, impenetrable, moveable Particles ... even so very hard as never to wear or break in pieces; no ordinary Power being able to divide what God himself made one in the first Creation."[2]

In the early 19th century, the English chemist John Dalton proposed that all the various kinds of materials are composed of pure, unalterable substances, the atomic elements, which cannot be divided by chemical means.[3] Any material object can be decomposed into these simple, elemental atoms that remain exactly identical wherever and whenever they are found. Nowadays, these elements are specified by atomic numbers, with hydrogen the lightest element numbered 1, through carbon 6 and oxygen 8, and up to gold 79 and beyond.

Elemental atoms are combined and bound together in thousands and millions of ways to form *molecules*, from the Latin for "little mass," and a countless number of them are hidden in everything we see. Every time you breathe, you take in roughly one million, billion, billion, or 10^{24}, molecules of oxygen from the air. A drop of water contains a comparable number of atoms, close to the number of stars in the Universe.

The tiny elemental atoms are much smaller than a speck of dust, and as invisible as germs, a phantom, or a spirit. As the Little Prince learned, "what is essential is invisible to the eye; the house, the stars, the desert — what gives them beauty is something that is invisible."[4]

Radioactivity

The most abundant atoms found on the Earth are exceptionally durable, and have been around for billions of years. Only relatively rare, naturally occurring elements are temporary and unstable, like uranium and other radioactive elements. Their discovery began at the close of the 19th century, at the Sorbonne in Paris, when Henri Becquerel found that minerals containing uranium emitted strong, invisible radiation that fogged photographic plates.[5] This mysterious, penetrating emission was being sent out at a regular pace in intense light or pitch darkness and at high or low temperatures. It eventually became known as *radioactivity*, with the term *radio* implying "radiation," so the radioactive atoms were radiation-active. No one knew just what the imperceptible rays were, how they were energized, or why the radioactive materials kept pouring out energy, seemingly nonstop.

Hearing of Becquerel's discovery, Pierre Curie, also a Professor of Physics at the Sorbonne, and the young graduate student he had recently married, Manya (Marie) Sklodowska Curie, began to investigate the new kind of rays.[6] Marie Curie developed methods of measuring the amounts being released, and she found that impure uranium ores emitted more rays than could be explained in terms of the uranium they contained.

The couple began a laborious two-year search for the unknown substance that was emitting the powerful rays. From one ton of uranium ore known as pitchblende they extracted just a few grams of powerful new radioactive elements that had not been previously known. One of them, called *radium*, released an energy that surpassed anything that had been achieved by chemical reactions. Crystals containing radium would light up an otherwise dark room, and also burn the skin, as Pierre Curie discovered to his dismay.

These findings became an international sensation. In 1903, the Nobel Prize in Physics was, for example, awarded jointly to Bequerel, for his discovery of spontaneous radioactivity, and to Pierre and Marie Curie for their joint researches on this radiation phenomenon. [Marie Curie was the first woman to win a Nobel Prize, the first person to win twice (in 1911 for Chemistry), and the first woman to become a Professor at the University of Paris. She died in 1934 due to illness brought on by exposure to radium and x-rays.]

The emanations of radium were investigated in greater detail by the New-Zealand born physicist Ernest Rutherford and the English radio-chemist Frederick Soddy while they were working at the McGill University in Montreal, Canada. They found that the radioactive rays included at least two distinct types, termed α rays and β rays, which are not waves of radiation but instead beams of energetic, fast-moving particles.[7] By using electric and magnetic fields, the two types of particles could eventually be separated and their physical properties determined. [8]

The rocks and soil under our feet are still radioactive billions of years after the Earth formed, and they still heat the interior of the planet.[9] So the radioactive atoms have been around for a very long time, and are just slowly falling apart. That's because it is ever so hard for a α particle to overcome the forces holding it within an atom.

The young Russian physicist George Gamow explained the mystery in 1928, while at the University of Göttingen, in what is now Germany. He proposed that the α particles have a range of possible behaviors with varying probability. Although they usually lack the energy to overcome the forces holding them within a radioactive atom, some of the α particles have a small non-zero possibility of penetrating that barrier and escaping to the outside world.[10]

In this surreal world of sub-atomic probability, one could relentlessly throw a ball against a wall, watching it bounce back countless times, until eventually the ball would tunnel through the wall, breaking out to the other side. As Ahab stated in Herman Melville's *Moby Dick*: "How can the prisoner reach outside except by thrusting through the wall."[11]

The known, steady decay rates of radioactive atoms have been used to clock the Earth's age. By combining these rates with measurements of the relative amounts of radioactive parent elements and stable, non-radioactive daughters, an age of about 4.6 billion years has been determined for the Earth.[12]

All the atoms, from the durable ones to the radioactive ones, are in a state of continual motion, with speeds that depend on their temperature.

Ever Moving Atoms and Molecules

Atoms and molecules are always in motion, pushing and jostling each other, coming together and pulling apart, in constant invisible turmoil. This incessant movement can be noticed if you add some wine to a glass of water; the two liquids do not remain separated but dissolve and move into each other. A similar thing happens when two gases are mixed together, as when smoke from a chimney moves out Into the air and disperses and disappears within it.

Molecules are always moving about because they are hot. That is, their energy of movement, and their state of agitation, increases with the temperature. The higher the temperature the faster a molecule moves, and at a specific temperature, lighter molecules tend to move faster than heavier ones. When the temperature drops to absolute zero, every molecule will cease to move, becoming totally immobile and completely at rest.

How fast and far do molecules move? Each molecule in the air of your room is moving with speeds of about 500 meters per second. That is about as fast as a rifle bullet moves when fired. But no molecule ever travels very far. There are so many molecules in the air that a given molecule hits another one after traveling only one ten-millionth of a meter. In just one second, this

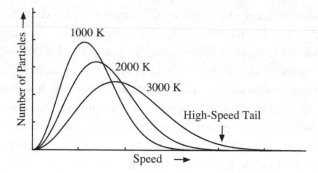

Figure 3.1   Speed distribution The speeds of particles with the same mass and three different temperatures. The peak of this distribution shifts to higher speeds at higher temperatures. The peak also shifts to higher speeds at lower mass when the temperature is unchanged.

molecule collides with other ones and is deflected from its course five billion times.

It was James Clerk Maxwell, from Scotland, who in 1860 realized that gas molecules gain or lose speed by collisions with one another. He showed that these encounters produce a statistical distribution of speeds, in which all the speeds might occur with a different and known probability.[13] At any given instant, the speed of most of the molecules in a gas is very close to the average value that increases with the temperature, but there is always a small percentage that move faster or slower than the average and this range also increases with the temperature (Fig. 3.1).

Maxwell's discovery was a true moment of unexpected insight, made at a time when no one else even imagined that a gas is composed of molecules whose motions determine its physical properties. In addition to atoms or molecules, his distribution can be used to describe the speeds of numerous sub-atomic particles inside the Sun. It applies to the motions of many different kinds of particles, as long as there are a lot of them in a state of thermal equilibrium, characterized by a single temperature.

As a result of the frequent collisions between each other, molecules are in haphazard movement and no molecule ever takes a direct path anywhere. They move about in a disoriented and random way, first moving in one direction and then in another one. This meandering journey can be observed by watching small particles or grains suspended in water.[14]

The observed motion does not originate either in the particles themselves or from currents in the water, but instead arises from water molecules that relentlessly strike nearby pollen particles that travel on a lively, disoriented

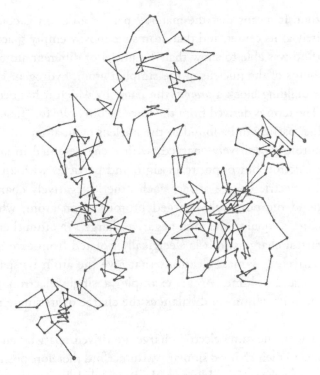

Figure 3.2 **Brownian motion** The irregular movement of a microscopic grain. [Adapted from Jean Perrin's *Les Atomes*, published in 1913].

path (Fig. 3.2). Watching the movement caused by the unseen molecular collisions is somewhat like inferring the motion of sea waves by observing the rocking motion of a ship.

The particles deep within a star are in a heightened state of movement due to the exceptionally high temperature down there. These atoms are so hot, and the collisions between them so frequent, that they are torn into their internal parts.

Sub-Atomic Particles

When bombarding gold leaf with beams of α particles in the second decade of the 20th century, Ernest Rutherford found that most of the α particles went right through the material as if it was not there.[15] To his amazement, only about 1 in 20,000 of them bounced right back from where they had come from, while all the others passed through the gold. This indicated that the mass of an atom is concentrated in a nucleus that is 100,000 times smaller than the atom. It meant that the material part of an atom is concentrated almost entirely at its center, and that atoms are mostly empty space.

Rutherford was able to show that the nuclei of different atoms contain various amounts of the nucleus of the simplest atom, hydrogen. He named this nuclear building block a *proton*, the name by which it has been known ever since. The term is derived from the Greek word *proto* for "first," because it was the first particle to be found in the nucleus of an atom.[16]

The proton is positively charged, with a charge equal in amount to that of the electron but opposite in sign, and particles with an opposite sign to their electric charge attract each other. Negatively charged electrons surround the positively charged protons in an atom, whose total positive charge is equal to the total negative charge. An atom therefore has no net electrical charge and it is electrically isolated from external space. It is the relatively low-mass electrons that give the atom its space-filling extendedness and volume. As an example, a single electron gives the hydrogen atom its volume and balances the charge of its single proton at the center.

Particles with the same electric charge are driven apart by an electrical repulsion, due to their charged similarity. Rutherford therefore postulated the existence of an uncharged nuclear particle, later called the *neutron*, to act as a sort of glue to help hold protons together in the atomic nucleus, and keep the protons from dispersing as they repelled each other (Fig. 3.3). As the name implies, the neutron has no electric charge. [Rutherford became director of the Cavendish Laboratory at the University of Cambridge in 1919, and under his leadership the English physicist James Chadwick discovered the neutron in 1932 — after an eleven-year search.[17]]

The proton and neutron have about the same mass, which is nearly 2,000 times that of the electron. But the mass of an atomic nucleus is always just a bit less than the sum of the masses of its protons and neutrons because they have expended energy in order to bind themselves together. That is, energy has been set free in the formation of a nucleus, and the mass defect between the nucleus and the combined mass of its components is a measure of how much energy was lost in the creation of the atomic nucleus.

Once astronomers knew about the internal ingredients of atoms, they could estimate how hot it is inside the Sun, and how fast the sub-atomic particles are moving within it.

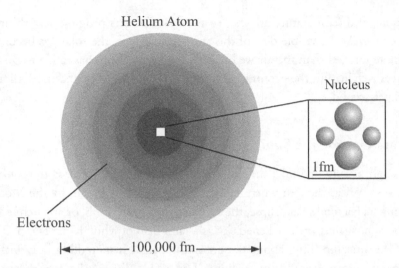

Figure 3.3 **Helium** The helium atom contains two electrons that swarm about the atom's nuclear center in largely empty space. The atom is about 100,000 times bigger than its nucleus, which consists of two protons and two neutrons.

What's Inside the Sun?

The American astronomer Jonathan Homer Lane first realized, in 1870, that the Sun would have to be very hot inside to support its enormous mass and retain its exceptionally large shape.[18] Hot particles, which we now know are protons, move about rapidly and frequently collide with each other to produce the gas pressure that keeps the Sun from collapsing under its great mass. [Since the mass of the electron is negligible compared to the mass of the proton, the mass of the Sun is provided by the protons, which determine the interior particle number density, the central temperature and the central gas pressure of the Sun.]

When the temperature rises to a central value of 15.6 million degrees kelvin, equilibrium is reached between the outward pressure of moving protons and the inward gravitational pull at the Sun's center. At this temperature, a proton within the Sun will be moving, on average, at a speed of about 500 kilometers per second.

In every layer within the Sun or any other star, the weight of the overlying gas must be equal to the outward-pushing pressure; otherwise the star would expand or contract, which is usually not observed. At greater distances from the center, there is less overlying material to support and the compression, pressure, and temperature are less, so the material gets progressively thinner and cooler. At the visible disk of the Sun, for example, the solar gas becomes far more rarefied than the air we breathe, the temperature has fallen to 5,780 degrees Kelvin, and the protons move at an average speed of about 10 kilometers per second.

What's Outside the Sun?

Although usually invisible, the outer solar atmosphere, known as the *corona*, can be seen near the Sun when its light is blocked, or eclipsed, by the Moon. During such a total solar eclipse, the corona is seen at the limb, or apparent edge of the Sun, against the blackened sky as a faint halo of white light (Fig. 3.4).

The amazing thing about the corona is that it is incredibly hot, with a temperature of a few million Kelvins. The visible disk of the Sun is several hundred times cooler than the overlying corona, which was entirely unexpected. It violates common sense, as well as the second law of thermodynamics, which implies that heat cannot be continuously transferred from a cooler

Figure 3.4   Eclipse corona streamers The million-degree solar atmosphere, known as the *corona*, is seen around the black disk of the Moon during a total solar eclipse on July 11, 1991. (Courtesy of the High Altitude Observatory, National Center for Atmospheric Research.)

to a warmer body without doing work. When we sit far away from a fire, for example, it warms us less. Contemporary solar astronomers attribute the heat of the corona to the interaction of magnetic fields that are always coming out from, and going into, the interior of the Sun.

The corona is so intensely hot that it cannot be entirely constrained by either the Sun's inward gravitational pull or by its magnetic forces. Some of the corona's solar electrons and protons escape into surrounding space, which means that the Sun's radiation is not all that moves past the planets. An eternal solar wind of electrons and protons is forever blowing the outer solar atmosphere away in all directions. Some of it sweeps past the Earth and engulfs it, which means that we live inside the Sun.

Although the Sun is continuously blowing itself away, the outflow can continue for many billions of years without significantly reducing the Sun's mass. Every second, the solar wind blows away a million tons, or a billion kilograms. That sounds like a lot of mass lost, but it is four times less than the amount of mass consumed every second during the nuclear reactions that make the Sun shine. This nuclear fuel is expected to last another 7 billion years, and that's how long the Earth will continue to be illuminated and warmed by the Sun.

This brings us to the discovery of how sunlight and other kinds of radiation move through space.

4. How Light Moves Through Space and Interacts with Matter

"For the invisible things of Him
from the creation of the world are clearly seen."

The Holy Bible, Romans (56 A.D.)[1]

Michael Faraday, a Devout Sandemanian

Michael Faraday was born on September 22, 1791, into an impoverished London family where there was barely enough to eat. He was withdrawn from school at 14 years of age, when he became an apprentice to a book-binder and bookseller. That turned out to be a blessing in disguise, for Faraday educated himself by reading the books that were being bound.

At the age of 20, and the end of his apprenticeship, Faraday was given a ticket to lectures by the eminent chemist Humphry Davy at the Royal Institution in London. Faraday sent a bound copy of the lecture notes he had taken to Davy asking for a job, and about a year later he was hired as an assistant in chemical experiments that were carried out in the basement of the Royal Institution. Davy and Faraday discovered that the chemical properties of substances are associated with their electrical powers, and that they are held together by positive and negative charges of electricity.

In 1821, Faraday married Sarah Barnard, after a two-year courtship and many love letters. They moved into modest rooms above the Royal Institution, where Faraday continued with his basement experiments. Michael and Sarah met at the Sandeman Church in London, an offshoot of the Church of Scotland, which provided them with spiritual sustenance throughout their lives.

The Sandemanian's worked hard, lived simply, and spoke plainly. They had no clergy, leader or social hierarchy and sought guidance from mutual knowledge, friendship, and God's word in the *Bible*. Ambition, pride, and the quest for worldly wealth were not, they thought, relevant to the true Christian who sought the eternal spiritual world rather than temporary and mundane involvement in politics, commerce or fashion. They insisted on the spiritual equality of each member irrespective of wealth, age, sex, or accomplishments.[2]

The Sandemanians took their commitments very seriously. Faraday, for example, was briefly excluded from their fellowship for missing a weekly service, which brought him shame and poor health and spirit. He was just barely excused after explaining that he was following the Queen's command to dine at Windsor Castle.

To Faraday, scientists had an obligation to share their understandings with everyone. He therefore gave yearly public Christmas lectures at the Royal Institution on topics such as chemistry, electricity and attractive forces. The chemical history of the candle is still a favorite. In another well-known lecture on Mental Education he mentioned in 1854 that: "The book of Nature, which we have to read, is written by the finger of God."[3]

In these widely appreciated lectures, Faraday conveyed the wonder, joy and excitement in the contemplation of Nature. He ignited the flame of curiosity in all kinds of people, from common laborers to Prince Albert, and opened the wide eyes of youth even further then normal.

Faraday viewed rewards, patronage and politics as undermining the purity of science. He thought that they were inconsistent with scientific objectives and the pursuit of truth. He twice refused the Presidency of the Royal Society when it was offered because it might require the political intrigue that he deplored. He also turned down a knighthood because he thought the English honors system was corrupt. To his dying days, he was exceptionally humble and preferred to remain plain Michael Faraday.

Faraday's Unseen Fields of Force

Faraday dismissed both mathematics and theoretical hypothesis as methods of understanding the physical world. To him, mathematical theories were no more than dishonest confessions of our ignorance. The only route to understanding Nature was through experimentally determined facts.

His experiments in electricity and magnetism were stimulated by the investigations of the Danish scientist Hans Christian Ørsted, who showed that electric and magnetic forces are interrelated. He noticed that the magnetic needle of a compass moved when direct current flowed through a nearby wire.[4] The magnetic needle also turned in opposite directions when placed above or below the current-carrying wire.

On reading of Ørsted's results, Michael Faraday and Humphry Davy immediately repeated and extended his experiments. They wanted to learn more about the invisible forces that reached out through empty space to connect the wire and compass needle, as if they were touching each other.

They confirmed that electricity can generate magnetic forces that pass through space unseen, and about ten years later Faraday discovered that a moving magnet can produce invisible electrical forces.[5] These two interactions are now used to make electric moters and to generate electricity.

Michael Faraday proposed that every electrically charged body is surrounded by an unseen electric field of force, and that every magnet is enveloped by an invisble magnetic force field. Both the electric and magnetic field of force could be viewed as a manifestations of a single unseen field of force, termed the *electromagnetic field*.

These invisible fields were supposed to permeate all of space, and their unseen lines of force were believed to stretch through all that exists. For Faraday, and Einstein after him, fields are a force underlying all discernable things. The material things we notice around us are just a limited perception of an unseen reality described by these fields. The points where these fields meet and focus are the points where we perceive matter to exist. The entire Universe is crisscrossed by the invisible lines of force, but the closer you are to the source of a field, the greater its power and hidden ability to act.

Faraday believed that God placed the invisible fields into the physical world at the time of its Creation, and that we can determine the laws by which their unseen powers act. These divinely ordained laws were supposed to govern the physical Universe, and the entire substantive world was believed to be at play in the fields of the Lord.

Ray Vibrations

In the early 1840s Faraday suffered an extended illness and a nervous breakdown. He felt sure that his days of discovering the hidden secrets of Nature were over. He was wrong. One of his most prophetic insights was made on April 3, 1846 during one of the Royal Institution's Friday Evening Discourses.

The chosen speaker, the distinguished Charles Wheatstone, panicked and left just before he was to speak. Faraday filled in by making some private and unpublished "Thoughts about Ray-vibrations." He attributed radiation and radiant phenomena to vibrations in the electromagnetic fields of force. When disturbed, they would vibrate laterally and send waves of energy along their lengths. Light, he suggested, was a manifestation of these field vibrations. He stressed that they were vibrations of the fields of force themselves, and not of any hypothetical luminiferous aether that some thought necessary for light to propagate in.

Although these were speculative thoughts, just vague impressions of Faraday's mind that appeared "only as the shadow of a speculation,"[6] the concept of invisible vibrations served as a guide to James Clerk Maxwell, whose equations described Faraday's results and tell us how light moves.

James Clerk Maxwell, by God's Grace

Maxwell was born in Edinburgh on June 13, 1831 as a member of the Clerk family that had been prominent in Scotland for two centuries. His father, John Clerk Maxwell, added the name Maxwell to that of Clerk as a condition for inheriting a 1500-acre estate on which he built the family house, named Glenlair. James spent his infancy and early boyhood there, but in his ninth year his mother died of abdominal cancer and he was sent to the Edinburgh Academy.

He spent six years at the Academy, which were followed by three years at the University of Edinburgh. Maxwell then proceeded in 1850 to the University of Cambridge, where he became a Fellow at Trinity College. While at Cambridge, Maxwell became seriously ill and was cared for by the Reverend C. B. Taylor and his family. The experience gave him a new perception of the love of God. When subsequently writing to Taylor, on July 8, 1853, Maxwell commented on his own sins and God's grace and guidance, and stated:

"All the evil influences that I can trace have been internal and not external, you know what I mean — that I have the capacity of being more wicked than any example that man could set me, and that if I escape, it is only by

God's grace helping me to get rid of myself.... by committing myself to God as an instrument of his will."[7]

Maxwell's belief in God played an important role in both his personal life and his work as a scientist. Shortly before his marriage to Katherine Mary Dewar in 1856, the devout Maxwell wrote to her about his strong belief in God, including: "Let us bless God even now for what He has made us capable of, and try not to shut out His spirit from working freely."[8]

Maxwell ceased to be a Fellow of Trinity College, whose fellowship excluded married men and all women. He was appointed Professor of Natural Philosophy at the Marischal College, Aberdeen. But he also lost his Professorship a few years later when Marischal was joined with another college into one university; the number of professors was reduced, and it was noticed that Maxwell was an unsuccessful teacher with a disconnected, rambling lecture style. [Astronomers and other scientists were known as *natural philosophers* until the late 19th and early 20th century when the term *scientist* was coined and came into use.]

There was no need for Maxwell to be concerned. He was soon appointed to a similar Professorship at the Kings College in London where he stayed from 1860 to 1865. During this time he explained the way we see colors, described physical lines of force, and delineated the kinetic theory of gases. In 1865 he also published his dynamical theory of the electromagnetic field, which predicted the existence of electric and magnetic waves of energy that travel through space unseen. These waves, he supposed, are produced by the changing fields of force that had been carefully studied by Michael Faraday. This explanation also indicated that visible light amounts to only a small band of the possible electromagnetic waves in space, all traveling at the same speed but with different wavelengths.

In 1865 Maxwell retired to Glenlair, but he was called out of retirement six years later to take a new Professorship of Heat, Electricity and Magnetism at the University of Cambridge. He became the first Director of the Cavendish Laboratory, which opened in 1874 and continues today as a place where scientific experiments and measurements are carried out. Upon becoming its Director, Maxwell stated that: "Those aspirations after accuracy in measurement, truth in statement, and justice in action are ours because they are essential constituents of the image of Him who in the beginning created not only the Heaven and the Earth but the materials of which Heaven and Earth consist."[9]

Maxwell died on November 5, 1879, at the age of forty-nine, of the same abdominal cancer that had killed his mother at about the same age. And near the end of his life, he told a friend that: "What is done by what I call myself is, I feel, done by something greater than myself in me."[10]

Maxwell's Equations and the Discovery of Radio Waves

Like Faraday, Maxwell believed that God made the Universe; that the laws that describe Nature are God's laws, and that every discovery further reveals God's great design. As a devout Christian, Maxwell also believed that God's word could be found in the *Bible*.

In 1865, Maxwell discovered how Faraday's invisible electric and magnetic fields work.[11] Every time a magnet moves, or an electric current varies, a wave of energy moves outward at the speed of light. Moreover, a time-changing electric field generates a magnetic one, and a time-varying magnetic field makes an electric one. This pumping cycle, from one kind of field to another, sends electromagnetic waves spreading through space.

To our delight, we now know that Maxwell's description also includes invisible electromagnetic waves that had never been imagined before. But no one realized these implications in Maxwell's time, and in fact there didn't seem to be anyone who understood his equations. Even Faraday wrote him to ask that he change his mathematical hieroglyphics to something he might grasp.

An exception was a self-taught Englishman, Oliver Heaviside, who slowly worked his way through Maxwell's complex mathematics, grasped their meaning, and simplified his maze of currents, displacements, inductions, symbols, potentials and forces to just four equations that describe the rates of change of the electric and magnetic field in space and time.[12] These are the four famous "Maxwell's equations" that every physics students learns.

A quarter of a century later, the experimental physicist Heinrich Hertz verified these equations by producing and detecting what we would now call radio waves in his laboratory at the Technische Hochschule in Karlsruhe, Germany.[13] He used a high-voltage, oscillating spark to generate the unseen waves that were detected with a loop of wire on the other side of the room. The waves had a wavelength of about one meter and moved at the speed of light, or about 299.79 million meters per second.

A young Italian entrepreneur, Guglielmo Marconi, read of Hertz's work, and realized that the Hertzian waves might be used to communicate across vast distances without the aid of connecting wires. In 1895, at the relatively young age of 21 years, he used a Hertz oscillator, or spark producer, to send wireless signals over a few kilometers at his father's country estate near Bologna, Italy. He established communication across the English Channel between England and France four years later and between England and Newfoundland two years after that.[14] In ensuing years, Marconi had a grand time sending communications from his yacht *Electra* while cruising in the Atlantic and Mediterranean and then from the Vatican City to the Pope's summer residence at Castel Gandolfo. He also enjoyed hunting, cycling, motoring and was twice married to women with aristocratic pedigrees.

Marconi became an international hero, established the American Marconi Company, which later evolved into the Radio Corporation of America, abbreviated RCA, and in 1909 received the Nobel Prize in Physics, jointly with the German physicist Karl Ferdinand Braun for their development of wireless telegraphy.

The pioneering investigations of Maxwell, Hertz, and Marconi provided a foundation for the exploration of the invisible Universe, which can be observed at wavelengths that are longer or shorter than those of visible light (Fig. 4.1). The long ones that Hertz and Marconi investigated are radio waves that can pass through clouds on a stormy day. Like ghosts, they even move through the walls of your house. Invisible x-rays have much shorter wavelengths than light. They can penetrate skin and muscle to reveal your bones.

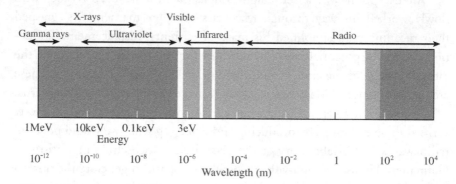

Figure 4.1   Electromagnetic spectrum Cosmic radiation only penetrates to the Earth's surface as the light we see with our eyes and as radio waves, respectively represented by the narrow and broad white areas. Radiation at other wavelengths has to be observed from above the atmosphere in Earth-orbiting satellites.

The use of invisible electromagnetic waves has changed our daily lives, from cell phones and global positioning systems to microwave ovens, radios, televisions, radar, and satellite weather images. Every star and galaxy in the Universe is also now emitting a host of these invisible electromagnetic waves, which all move at the speed of light.

The Speed of Light

It was not until the 17th century that astronomers discovered that light does not move instantaneously through space. The Danish astronomer Ole Rømer and Giovanni Domenico Cassini, Director of the Royal Observatory in Paris, noticed a varying time between eclipses of Jupiter's large inner moon Io, of up to 22 minutes variation in an orbital period of about 42 hours.[15] Both astronomers concluded that it was not the orbit of Io around Jupiter that changed, but the time it took light to travel from Io to the Earth, which depended on the Earth's position in its orbit around the Sun. When the Earth was on the side of its orbit that is closest to Jupiter, the observed eclipse period for Io was shortest, and when the Earth was on the opposite side of its annual orbit around the Sun, Io's apparent eclipse period was longest.

This meant that light does moves at a definite speed. Rømer and Cassini neglected to specify that speed, which would have been equal to the diameter of the Earth's orbit divided by the time difference between the longest and shortest observed Io periods. However, the distance between the Earth and the Sun was not then accurately known. Using today's estimates for this distance, their measurements would correspond to a light speed of about 227,000 kilometers per second.

The American physicist Albert Abraham Michelson became the recognized expert in designing and refining instruments for use in precise determinations of the velocity of light, as well as other astronomical measurements. Michelson was born on December 19, 1852 in Strzelno, Prussia (now Poland) and traveled to northern California with his family while still a child. He grew up in the rough gold-mining settlement of Murphy's Camp, where arguments were usually settled with fists, knives or bullets. When Albert reached the age of 12, his parents sent him for high school in San Francisco, and they moved to a silver-prospecting center in Virginia City, Nevada.

After graduation from high school, Albert took a competitive examination to enter the United States Naval Academy in Annapolis, Maryland, and eventually received an appointment there by President Ulysses S. Grant, graduating in 1873 with just less than a third of the entering class. Following graduation, Ensign Michelson began a four-year term as instructor in physics and chemistry at Annapolis, where he started two decades of precision measurements of the speed of light. The transmission of light between mountain tops in 1926, led to his final figure of 299,796 ± 4 kilometers per second, quite close to today's accepted value of c = 299,792.458.[16]

The study of light inspired a "pleasure, satisfaction, almost a reverence" in Michelson. He was fascinated by the rainbow of color produced by prismatic raindrops, at the beauties of coloring found on beetles, butterflies and hummingbirds, and at "the intricate wonders of symmetrical forms and combinations of forms, which are encountered at every turn." When trying to explain it all to his daughter, he said "it doesn't matter if you don't understand it now as long as you realize the wonder of it."[17]

Michelson resigned from the Navy in 1881, and the next year began teaching at the Case School of Applied Science in Cleveland, when he became intrigued by the possible motion of the Earth through the stationary aether, and devised a way to measure this movement. It was then thought that light waves must propagate in something, named the aether, just as sound waves are carried in air. Many scientists, from Newton to Maxwell, had firmly believed in such a light-carrying aether for nearly two centuries.[18]

Michelson's plan was to project a beam of light in the direction in which the Earth is traveling in its orbit, and one at right angles to this. The first beam, he thought, would naturally be retarded by the flow of aether passing the Earth. The second beam, crossing this current at right angles, should arrive ahead of the first by a length of time determined by the velocity of the Earth through the aether.

In the instrument that Michelson designed to make the measurements, a beam of light is split into two parts moving at right angles to one another; when reflected back and recombined, they produce an interference pattern, in which the light wave crests either added together or cancelled — hence the instrument's name *interference-meter* that is abbreviated *interferometer*. Such an interferometer can be used to accurately measure any changes in the speed or path lengths of the two beams.

Michelson's idea was to use the interferometer to see if light travels with the same velocity in all directions. That might measure how the speed of light depends on the Earth's motion through the stationary aether. If there is an aether, then the speed of light would vary when moving into the aether or against it, like a boat sailing with or against the wind or a swimmer moving downstream or struggling upstream.

Michelson's first attempt at using his interferometer to measure the relative velocity of the Earth and the aether, in 1881, indicated to him that there was no such motion and that "the hypothesis of a stationary aether is thus shown to be incorrect."[19] It was not until 1886 and 1887 that Michelson and his friend, the chemist Edward W. Morley of neighboring Western Reserve University, repeated this experiment with greater care and refinement. They were both dedicated to accuracy and precision in measurements, and their precise determination of the speed of light suggested that the aether does not even exist.

Their interferometer was set up in a basement laboratory and mounted on a massive sandstone slab that floated on a pool of mercury to remove unwanted vibrations. Light beams were then sent through the interferometer into the supposed aether wind, in the direction of the Earth's motion, and at right angles to this path. After a brief interruption by Michelson's nervous breakdown, the experiment resulted in an unexpected result. Michelson and Morley found that there was no detectable difference in the interference pattern produced when a beam of light was sent into the aether wind in the direction of the Earth's motion or directed at right angles to it.[20] Moreover, there was no difference in the measured speed of light when the Earth was traveling toward the Sun and away from it half a year later.

At this time, Michelson experienced a lot of personal difficulties. In 1887, a maid charged him with assault and battery; a court found him innocent of these charges. About a decade later, his wife of more than twenty years divorced him, and he seems to have remained a lonely person despite taking a former student as a second wife. Albert always possessed an astonishing indifference to family life and to people in general, confided in no one including either his wife or his children, and was rarely moved by the human attributes of ambition, envy, fate, and love.[21]

In 1889, Michelson left Case for Clark University of Massachusetts, and in 1893 moved to the University of Chicago, where he headed the Physics

Department until 1929. During his tenure at Chicago, Michelson was awarded the 1907 Nobel Prize in Physics "for his optical precision instruments and the spectroscopic and metrological investigations carried out with their aid," the first American to receive the award.

The Michelson-Morley experiment indicated that the velocity of light is constant, exactly the same in all directions and at all seasons, and independent of the motion of the observer. Light was no longer viewed as luminous ripples in the mysterious aether, and radiation was supposed to always propagate with a definite, unchanging speed in empty space.

This discovery played an important role in Einstein's famous 1905 account of moving bodies whose physical properties are relative and can change in high-speed motion.[22] He assumed that the velocity of light in empty space is independent of the motion of its emitting source and the motion of any observer, which means that light's speed is the same for any star or galaxy and for all observers, wherever they might be. Moreover, in his theory light has a special, limiting speed, the fastest that any material object can travel.

Nothing in our material world outruns light; it is the fastest thing around. That is because an object's mass increases without bound when it approaches the speed of light, and there is nothing that can propel such a massive object so fast. It would take an infinite amount of energy to accelerate any material object to the speed of light.

Einstein was later taken to task for omitting any mention of Michelson and Morley in his influential Special Relativity paper, or for that matter related publications of Lorentz, FitzGerald and Poincaré. Not until 1931 did Einstein publicly honor Michelson for his experiment, addressing him in person with: "It was you who led physicists into new paths, and through your marvelous experimental work paved the way for the development of the theory of relativity. You uncovered an insidious defect in the aether theory of light as it then existed, and stimulated the ideas of H. A. Lorentz and FitzGerald out of which the *Special Theory of Relativity* developed."[23]

To astronomers, the important implication of the Michelson-Morley experiment is that starlight is always in motion in empty space at the definite and precise speed of light. Moreover, all electromagnetic waves, regardless of wavelength, move though the vacuum of space at the same constant speed of light. They can persist forever in empty space, never speeding up and never slowing down or coming to a rest. Once emitted, radiation from any star or galaxy might therefore travel for all time in vacuous space, bringing its message forward to the end of the Universe. When some of that radiation is intercepted at the Earth, it tells astronomers about the cosmic object back when it emitted its radiation, as it was then and not as it is now.

Sunlight requires about 8 minutes to travel from the Sun to the Earth, and it takes just a little more than 4 years for starlight to reach us from the nearest star other than the Sun. Moving at the speed of light, it takes 2.3 million years for light to travel from the nearest spiral galaxy, Andromeda, to the Earth, and the light observed from very distant galaxies was emitted about 13 billion years ago. In effect astronomers watch cosmic history race at us at the speed of light, and trace out changes over millions and billions of years.

We can even witness the youth of distant stars that existed before the Sun and Earth were formed, about 4.6 billion years ago. Some of the most distant galaxies may no longer exist, and they may have been embryonic galaxies when the light now reaching the Earth began its journey. These galaxies could have perished over time, but their light can survive without change.

Nevertheless, once radiation encounters matter, it no longer behaves like a wave. Max Planck showed that light then acts like a particle containing little packets of light energy.

Planck's Uncomfortable Life and Resolute Search

Max Karl Ernst Ludwig Planck, or Max Planck for short, was born on April 23, 1858 at Kiel, Germany within an academic family. Both his grandfather and great-grandfather were Professors of Theology in Göttingen; his father was a distinguished Jurist and Professor of Law at the University of Kiel.

When Max was nine years old, his father received an appointment at the University of Munich, and Planck attended the city's renowned Maximilians Gymnasium. After graduation in 1874, at age 17, he entered the University of Munich, and in 1877 spent a year of study with Hermann von Helmholtz and Gustav Kirchhoff at the University of Berlin. To Planck, their lectures were boring, monotonous and uninspiring. Returning to Munich, he received the doctoral degree in July 1879, at the unusually young age of 21. The following year, Planck completed his Habilitationsshrift (qualifying dissertation) on the equilibrium states of bodies at different temperatures, and became an unpaid Privatdozent (lecturer) while awaiting an academic position.

In 1885, with the help of his father's professional connections, he was appointed an Associate Professor at the University of Kiel. In 1889, after the death of Kirchhoff, Planck received an appointment at the University of Berlin, where he had a close personal and professional relationship with Helmholtz. In 1892, Planck was promoted to Full Professor, and he eventually became Dean at the University, which enabled him to establish a new Professorship there for Albert Einstein in 1914. Planck remained in Berlin for the rest of his active life, and died in Göttingen in 1947 at the age of 89.

From a young age, Planck was not only gifted in science but also in music. He had absolute pitch, took singing lessons, was a member of the singing club at the University of Munich, composed songs and operas, and was an excellent pianist. In later years he found pleasure in daily playing the piano, which may have provided calm distraction from the tensions of his unlucky personal life.

After a little more than a decade of marriage, Planck's first wife died in 1909, probably of tuberculosis. They had four children, two sons and twin daughters. In the First World War, his eldest son was killed and the French imprisoned the other son. During World War II, the Gestapo executed his youngest son for his part in an attempt to assassinate Hitler. One daughter died giving birth. The surviving twin fell in love with her sister's husband. They married and two years later she died in childbirth. And when he was 85 years of age, an Allied bomb destroyed Planck's house and everything in it, except Planck who was elsewhere. With stoic strength, he continued with his work, sustained by his faith in God.

Max Planck reasoned that both religion and science require a belief in God for their activities, and argued that the truth of their compatibility is the historic fact that the very greatest natural scientists of all times — men such as Kepler, Newton, Leibniz — were permeated by a profound religious attitude.

"Inner peace of mind and soul," he wrote, "is secured by a firm [and universal] link to God and by an unconditionally trusting faith in his omnipotence and benevolence." Our scientific understanding "similarly demands that we admit the existence of a real world independent from us, a world which we can never recognize directly, but only indirectly by our measurements."[23]

Planck noticed that definite laws govern the physical world of Nature, and he identified that world order with the God of religion. He believed that there are "everywhere active and mysterious forces" in Nature, and that a search for the laws that describe them is comparable to seeking the Divine. In other words, the goal of a scientist's quest is to approach God and His world order. In Planck's own words:

"Religion and natural science are fighting a joint battle in an incessant, never relaxing crusade against skepticism and against dogmatism, against disbelief and against superstition, and the rallying cry in this crusade has always been, and always will be: *On to God.*"[24]

Planck Invents Light Quanta

Planck's predecessor at the University of Berlin, Professor Gustav Kirchhoff, found in 1860 that the heat radiation of a black body, which absorbs and emits all radiation that falls on it, has an energy that depends only on the radiation wavelength and the temperature.[25] This became known as *Kirchhoff's law*. The black body radiation does not depend on the volume or shape of the source, or on the material it is made out of. This meant that the radiation is disconnected, a thing on its own, and when in space radiation is a reality independent of material bodies. This universality fascinated Planck, and he spent years trying to find a deeper, absolute explanation of it.

Planck was not alone in seeking the general properties of black body radiation. In 1879, the Austrian Joseph Stefan, at the University of Vienna, used experiments to show that the total power emitted from a heated body is proportional to the fourth power of the temperature.[26] And when size is taken into account, the power also varies as the square of the radius. This explains why giant stars are more luminous than the Sun. With his relation, Stefan also used observations of the Sun to determine the temperature of its visible disk, at about 5700 degrees kelvin.

In 1886, the American astronomer Samuel Pierpont Langley reported laboratory spectral measurements of the radiation from heated copper using the bolometer, an instrument he invented for detecting infrared radiation.[27] His observations displayed temperature-dependent intensity maxima and rapid drops in intensity with both increasing and decreasing wavelength (Fig. 4.2).

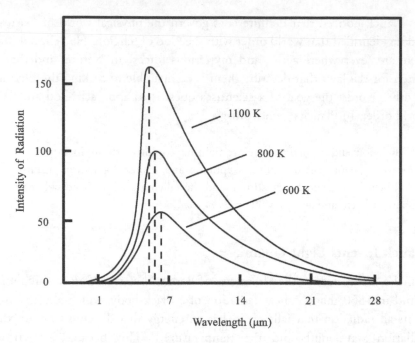

Figure 4.2   Hot emission The intensity of radiation from copper that has been heated to different temperatures and measured at infrared wavelengths in units of micrometers or 10^{-6} meters. [Adapted from S. P. Langley's article published in the *Philosophical Magazine* **21**, 394–409 (1886).]

As the temperature increases, more intense radiation is emitted at all wavelengths. Moreover, the wavelength of the most intense radiation shifts toward the shorter wavelengths when the temperature rises.

By the 1890s several investigators, both experimentalists and theorists, were trying to determine and describe the spectral distribution of black body radiation. Precise measurements of the radiation at various wavelengths were also being performed at the Physikalisch-Technische Reichsanstalt in Berlin, where Wilhelm Wien found that the radiation is most intense at a wavelength that is inversely proportional to the temperature.[28] He received the 1911 Nobel Prize for Physics "for his discoveries regarding the laws governing the radiation of heat."

Over a period of five years, Max Planck sought a way of rigorously deriving Wien's law and tried to find a more comprehensive explanation for the intensity of black body radiation as a function of its wavelength or frequency.

The German physicist Heinrich Hertz had discovered waves of invisible radiation in the previous decade, and this led to speculation that numerous "Hertzian oscillators" produce the heat radiation emitted by black bodies. Planck supposed that such microscopic oscillators might resonate at all possible wavelengths, and looked at ways their radiation might be distributed in energy amongst them.

With dogged perseverance, Planck continued in his search, despite several mistakes and setbacks when new experimental data were found inconsistent with his latest results. In what he described as "an act of desperation" and "a fortunate guess," he assumed that the total energy of all the black body oscillators in the cavity walls was distributed into finite portions of energy, each proportional to the oscillator frequency. Planck called the constant of proportionality the *quantum of action*, and it is now known as *Planck's constant*. Radiation at shorter wavelengths, or higher frequencies, has larger quantum energy and *vice versa*.

Using this approach Planck found, at the turn of the 20th century, an equation that described the spectral distribution of the energy emitted from a black body at any given temperature, which agreed with all the observations.[29] This spectral distribution for the black body radiation intensity peaks at a wavelength that is inversely proportional to the temperature, drops precipitously at shorter wavelengths, and falls off gradually at longer ones (Fig. 4.3).

The entire spectrum increases in intensity and shifts toward shorter wavelengths as the temperature increases. That's the reason that very hot, million-degree objects emit most of their radiation at short x-ray wavelengths, while cold interstellar space is most luminous at long radio wavelengths.

The Sun's radiation spectrum closely matches this distribution at a temperature of 5780 degrees kelvin (Fig. 4.4). This spectrum is most intense at visible wavelengths, which explains why the Sun illuminates the world we see. It is also the reason that we notice other stars, of comparable temperatures, when looking up at the dark night sky.

Planck's spectral distribution formula supposed that black body radiation does not interact with matter in a continuous stream at all possible energies, but instead in discrete bundles of energy, which Planck called *quanta*. That is, the radiation was not being expelled in a steady flow like water from a hose, but instead like microscopic bullets from a machine gun.

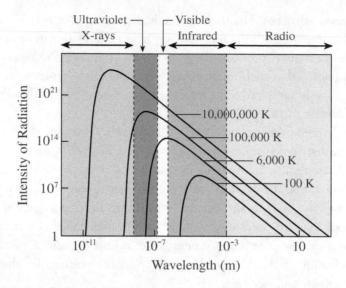

Figure 4.3   Black body radiation Intensity of black body radiation plotted as a function of wavelength in meters. At higher temperatures the wavelength of peak emission shifts to shorter wavelengths.

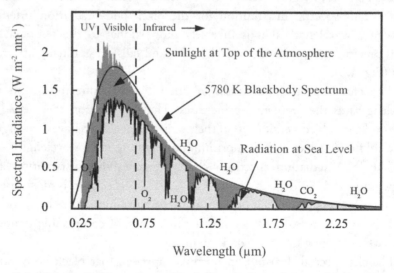

Figure 4.4   Solar radiation spectrum The Sun's radiation at the top of the Earth's atmosphere (*gray*) and below the atmosphere at sea level (*dark*). The maximum intensity occurs at visible wavelengths, and a black body at a temperature of 5780 degrees kelvin describes its spectral distribution. Oxygen, water vapor, and carbon dioxide molecules in the air absorb the sunlight. The solar irradiance is given in units of watts per square meter per nanometer, and the wavelength is in micrometers, or 10^{-6} meters.

The 1918 Nobel Prize in Physics was awarded to Max Planck for "his discovery of energy quanta."

How Radiation Interacts with Matter

In the meantime, back in 1905, Albert Einstein had interpreted the photoelectric effect in terms of the quantum interpretation of the interaction of light with matter.[30] When you shine ultraviolet light on a metal surface, some electrons will come out of the metal's atoms, which is known as the *photoelectric effect* (Fig. 4.5). It was found in the laboratory that the energy of these liberated electrons is unrelated to the intensity of the light, but simply related to its frequency, and that the total energy an electron absorbs from the light is exactly one quantum of energy. The product of the radiation wavelength and frequency is equal to the speed of light.

The energy transfer from one light quantum to a single electron is independent of the presence of other light quanta. Moreover, an atom only ejects an electron when the radiation frequency is above a threshold-frequency. Einstein received the 1921 Nobel Prize in Physics, "especially for his discovery of the law of the photoelectric effect."

The indestructible and indivisible light quanta are now known as *photons*, a term that was coined by the American chemist Gilbert N. Lewis in 1926.[31] Photons are created whenever a material object emits radiation, and the photons are consumed when matter absorbs radiation. The interaction of each

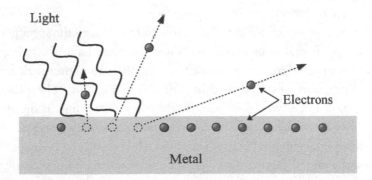

Figure 4.5   Photoelectric effect A metal can emit electrons when ultraviolet light shines on it. These electrons indicate that light behaves as a packet of energy, called a photon, when it interacts with matter.

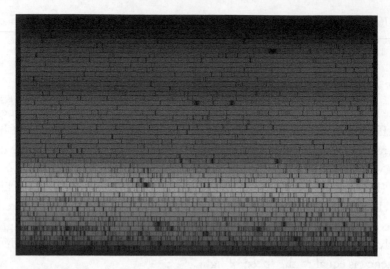

Figure 4.6   Solar absorption lines The visible portion of the Sun's radiation displayed as a function of wavelength. When we pass from long wavelengths to shorter ones (*left* to *right*, *top* to *bottom*), the spectrum ranges from red through orange, yellow, green, blue and violet. Dark gaps in the spectrum are due to absorption by atoms in the outer atmosphere of the Sun. (Courtesy N. A. Sharp, NOAO/NSO/Kitt Peak FTS/AURA/NSF.)

photon with matter depends on its energy, which is exactly the energy of Planck's quanta — the product of Planck's constant and the radiation frequency. This is the reason why energetic, high frequency x-rays pass through your skin and muscles with very little absorption until they reach your bones. Less-energetic ultraviolet sunlight can burn your skin, while visible sunlight, of even lower frequency, just warms your face.

The composition of any material can be inferred from the specific wavelengths at which it absorbs radiation. When the Sun's light, for example, is examined with sufficiently fine wavelength intervals, numerous dark features are found irregularly distributed within the colors (Fig. 4.6). These dark features are known as *spectral lines* because they look like a line in the spectral display. They are further designated as *dark lines* or *absorption lines* since atoms in a relatively cool tenuous gas produce them when absorbing the radiation of hotter, denser underlying material. When broad wavelength intervals are used, adjacent bright emission obscures the dark places that cannot then be found.

Just seven of the dark gaps of missing colors in the Sun's light were first noticed in 1802 by William Hyde Wollaston, a London chemist.[32] Hundreds of the dark spectral lines were subsequently discovered by the Bavarian glass-maker Joseph Fraunhofer, who had a rather grim early life. He was the eleventh child of an impoverished father, who died when the youngster was just 12 years old. Two years later, the slum building in which he lived collapsed, trapping the young Fraunhofer and killing everyone else inside. When the Bavarian prince, Maximilian Joseph, heard of the ordeal, he rescued Fraunhofer from his misery, giving him enough gold coins to follow an interest in optics, lens making, and the Sun.

By directing incoming sunlight through a slit, and then dispersing it with a prism, Fraunhofer was able to overcome the blurring of colors, and discovered numerous dark features in this spectral display.[33] He measured the wavelengths of these dark spectral lines, and catalogued more than 300 of them. Fraunhofer also found that bright stars exhibit spectral lines like those seen in the Sun, as well as some lines that he did not detect in sunlight. Unlike Wollaston, Fraunhofer was convinced that the spectral lines originate in the Sun or the other stars, but he did not know what caused the dark lines to appear in the stellar spectra, or where the missing colors went.

It was the German scientist Gustav Kirchhoff and his chemist colleague Robert Bunsen who showed that every chemical element, when burned and vaporized as a gas, emits brightly colored lines at unique wavelengths. Copper, for example, provides a green color to a flame, and sodium burns a bright yellow. These bright spectral features are known as *emission lines* because substances heated to incandescence emit them.

One evening, the two men noticed distant buildings that were burning, and detected emission line spectra in the flames. This suggested that they had a way of investigating the Sun's fires from a distance. By comparing the wavelengths of the Sun's dark absorption lines with those of emission lines from elements vaporized in their laboratory, Kirchhoff and Bunsen identified several chemical elements in the solar atmosphere.

As Bunsen wrote in 1859:

"At the moment I am occupied by an investigation with Kirchhoff, which does not allow us to sleep. Kirchhoff has made a totally unexpected discovery, inasmuch as he has found out the cause for the dark lines in the solar spectrum and can produce these lines artificially intensified both in the solar spectrum and in the continuous spectrum of a flame, their position being identical with that of Fraunhofer's lines. Hence the path is opened for the determination of the chemical composition of the Sun and the fixed stars."[34]

They had unlocked the chemistry of the Universe! Each chemical element, and only that element, produces a unique set, or pattern, of wavelengths at which the dark lines fall. It is as if every element has its own characteristic signature that can be used to identify it, as a fingerprint or DNA sample might identify a person. Every one of the numerous absorption lines found in the Sun's spectra have been identified with a specific chemical element or compound.

Some of the Sun's spectral lines are exceptionally dark, extracting great amounts of energy from sunlight. They are produced by hydrogen, sodium, magnesium, calcium and iron. It was therefore initially supposed that the Sun is made out of the same material as the Earth in similar abundance, but this is only partly true. Many of the visible solar lines were associated with hydrogen, which is by far the most abundant element in the Sun and most other stars but a terrestrially rare element.

Darker absorption lines generally indicate greater absorption and therefore larger amounts of the absorbing element, but the strength of an element's absorption lines depends only to some extent on an element's abundance. There are other mitigating circumstances, so unlocking the chemistry of the Universe was not as straightforward as scientists initially supposed.

Atoms, for example, exist in altered physical states at the high temperatures that prevail within stars, and this can result in a change in the intensity of the spectral lines that are observed in stellar atmospheres. Once this was understood, astronomers could estimate the number of atoms or ions responsible for the production of different dark lines within the Sun's colors. They found that there is a systematic decrease in the abundance of solar elements with increasing atomic number and weight. So the heavier elements are less abundant in the Sun than light ones.

This concludes our discussion of how light moves and interacts with matter, and brings us to the ways the stars move.

5. The Stars are Moving

"Silently, one by one,
In the infinite meadows of Heaven.
Blossomed the lovely stars,
The forget-me-nots of the angels."

Henry Wadsworth Longfellow (1847)[1]

How do the Stars Move?

It looks as if the stars are moving overhead each night, but these apparent movements are instead due to the Earth's rotation beneath the fixed stars. The turning Earth also explains why the Sun seems to rise and set each day. As the Earth rotates, day turns into night and the stars slide by.

Aside from the Earth's rotation, there is another more subtle motion of the stars that was discovered long ago, by the Greek astronomer Hipparchus around 150 BC. The appearance of a *nova stella*, a new star, at a place in the sky that no star had been seen before, inspired him to compile an accurate star catalogue of 850 bright stars. The listing of their brightness and position could, he thought, be used to determine if a star brightened or moved in later years.

When comparing his measurements of the stellar locations with those of his predecessors, Hipparchus discovered that some of these stars were apparently moving by the same amount and in the same direction as time went on. He suspected that all of the stars were slowly and steadily moving together, which Ptolemy confirmed about two hundred years later. The entire celestial sphere seemed to be twisting eastward at the rate of one angular degree per century, but this apparent shift in position is also due to the Earth's movement.

The Earth not only spins; it gyrates like a very large and slow, wobbling top. As shown by Isaac Newton, the gravitational pull of the Moon and the

Sun on the Earth's elongated shape causes a slow circular twist of the Earth's axis of rotation in space. It completes one circuit, every 26,000 years.[2]

So despite eons of stellar observation in antiquity, there was not a shred of evidence to contradict the belief that the stars are rooted in the sky. They always appeared at the same place in the celestial sphere, and never changed their apparent separations on it. That's why we can identify long-lived patterns amongst groups of stars, the constellations.

In contrast, it seems like everything on the Earth moves. Streams flow downhill, waves rise and fall on the sea, clouds drift across the blue sky, and tree leaves quiver in the wind. We all rise from bed to begin our daily movement across the land, and the entire planet spins on its axis and moves around the Sun.

And when you stop to think about it, the stars ought to move. Without motion, there would be nothing to keep the stars apart and suspended in space. Their mutual gravitation would eventually pull them into a single mass. There is not one star that is completely at rest.

We now realize that a star's motion in space manifests itself in two ways, depending on the method used to observe it (Fig. 5.1). One component of the motion is the "sideways" velocity directed perpendicular or transverse to the line of sight. The radial velocity is the other part of the motion. It is the component moving toward or away from us in the direction of the star. When

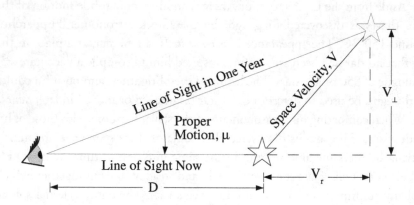

Figure 5.1   A star moves The transverse velocity of a star, $V_\perp$, which is perpendicular to the line of sight, can be inferred from measurements of a star's distance, D, and proper motion, μ. The radial velocity, V_r, directed along the line of sight, can be determined from the Doppler shift of the star's spectral lines. When these two velocity components are combined, the relative speed in space is obtained.

a star is moving straight at you, there is no perpendicular motion, and if the star is moving directly across your line of sight, the radial motion is reduced to zero. When both velocity components are known, we can determine the speed and direction of the star in three dimensions.

The Proper Motions of the Stars

The English astronomer Edmond Halley discovered that the stars do move on their own accord, but to do this he had to compare his observations with those made long before he was born. In 1718 he noticed that his determinations of the locations of a few bright stars differed from those measured by the Greek astronomer Hipparchus around 150 BC and recorded by Ptolemy in his *Almagest* in the second century.[3] So it took about 1,800 years before anyone noticed that a star could move in the sky.

In 1760 the German astronomer Johann Tobias Mayer fully confirmed Halley's discovery of stellar motion.[4] Mayer was not looking for stellar motion. He wanted to improve the way mariners and surveyors were using observations of stars to get their bearings, and to measure a person's location on the Earth. Since these measurements depended on the accuracy of the stellar observations, Mayer was investigating instrumental and atmospheric effects that could influence them. When he compared his own newly-acquired measurements of the positions of stars in the sky with those made by the Danish astronomer Ole Rømer only half a century before, Mayer found differences that could only be attributed to stellar movements.

The motion that Halley and Mayer detected is the "sideways" component of velocity directed perpendicular or transverse to the line of sight. It produces an angular change in position known as *proper motion*. The term suggests that the motion belongs to the star, and is proper in that regard. But the observed movement might nevertheless be attributed to either the star's transverse motion in one direction, the Sun's movement in the opposite one, or some combination of both motions.

When thinking about this uncertainty, Mayer noted that if the Sun was moving toward some region in space, all the stars which appear in that region would seem to be gradually separating from each other, while those in the opposite part of the sky would seem to be joining up. It is similar, he supposed, to walking through a forest in which the trees in front of you appear to move to your sides as you approach them, and those behind you seem to merge together. It wasn't until 1783 that William Herschel found such a pattern from the proper motions of just seven stars, and concluded that the Sun was moving toward the constellation Hercules.[5]

More than a century later, the Dutch astronomer Jacobus C. Kapteyn, of the University of Gröningen, found that nearby stars are apparently moving in two preferred directions.[6] They seemed to be traveling in a pair of large intermingled streams that pass through each other while moving in opposite directions in the Milky Way. The two star streams were found wherever Kapteyn and his colleagues looked, and they spent a lot of time looking. Altogether, the proper motions of some 2,400 stars were measured for Kapteyn's 1905 report on star steaming.

Telescopes lofted above the Earth's obscuring atmosphere have now been used to detect much smaller star-location changes than can be detected from the ground, with or without a telescope. Instruments aboard the *HIPPARCOS* satellite have determined the accurate positions and proper motions of more than 100 thousand stars.

Proper motion belongs to a star, but it isn't a velocity. It is the angular rate at which a star moves across the sky over the years or centuries, and it does not by itself determine the speed of motion. To convert a star's proper motion into a velocity, you have to know the star's distance, and no one knew the distance of any star other than the Sun until more than a century after Halley's proper-motion discovery.

Assuming that all stars move through space at roughly the same speed, those closest to Earth should display the largest proper motion over a given length of time. A nearby bird flying overhead similarly travels rapidly across a great angle, while a high-altitude one moving with the same speed barely creeps across the sky. That is how a duck hunter estimates the distance of a duck — by its angular speed.

So the nearest stars ought to exhibit the largest proper motion, provided they all move at about the same speed. That is why Wilhelm Bessel choose the "flying star" 61 Cygni to make the first measurement of a star's distance in 1838.[7] It had the largest proper motion known for a star at the time.

But lets now take a brief diversion and look at Halley's diverse interests.

A Glimpse at Edmond Halley's Interesting Life

Edmond Halley was born November 8, 1656 in the outskirts of London. He began his education at St. Paul's School, London, where he excelled in the study of classics and mathematics, and in the summer of 1673 he entered Queen's College at Oxford University. While at Oxford, Halley assisted John Flamsteed at the Royal Observatory, which was under construction at Greenwich with the support of King Charles II for the purposes of navigation by the ships of the seafaring nation.

Influenced by Flamsteed's compilation of the locations of northern stars in the sky, Halley proposed to do the same for the Southern Hemisphere. With financial assistance from his wealthy father and the King, and a letter of introduction to the East India Company, he sailed in 1676 to the South Atlantic island of Saint Helena, where he catalogued the accurate positions of 341 southern stars. When Halley returned to England in 1678, he published his observations,[8] was awarded a degree from Oxford University, and was elected a Fellow of the Royal Society at the age of 22.

During his sea voyage, Halley also mapped the prevailing trade winds over the oceans, and identified solar heating as the cause of atmospheric motions. He subsequently built and personally tested a diving bell for underwater exploration, traveled in an English war vessel to make magnetic charts of the Atlantic Ocean, and was one of the first to examine the statistics of mortality and age, which allowed the English government to sell life annuities at a price determined by the age of the purchaser. In other words, Edmond Halley was a curious and intelligent fellow who was captivated by the natural world and liked to find out how it worked.

Religion played an important part of Halley's life, as it did for just about everyone in England at the time. As an example, he was denied the post of Savilian Professor of Astronomy at the University of Oxford because he was not an orthodox Christian. But he was appointed to the similar Geometry Professorship after the death of his religious enemies, including the Archbishop of Canterbury.

Like Newton, Halley seems to have denied the strict equality of Christ and God, or the Son and the Father, and he similarly avoided strict interpretations of the *Bible*. As an example, Halley proposed to the Royal Society that

Figure 5.2   The eve of the deluge The arrival of a comet foretells the great flood in Noah's time. (John Martin made this painting in 1840; collection of Her Majesty the Queen.)

the great flood of Noah's time may have been due to a comet (Fig. 5.2), and perhaps not the direct result of God opening the flood gates of Heaven because of humanity's misdeeds.[9]

Nevertheless, Halley retained a strong faith in an all-wise and all-powerful God.[10] In the Latin *Ode* he prefixed to Newton's *Principia* at its publication, Halley acclaimed his achievement as an insight to God the Creator of the Universe. For both Halley and Newton, the goal of astronomy was to comprehend Nature, and to thereby gain access to the Divine and knowledge of God.[11]

To return to stellar motion, there is another component of their movement that is directed along the line of sight to them.

Radial Velocities of Stars

If a star is headed straight toward or away from you, its speed is as difficult to judge as that of an approaching car in the opposite lane of a highway. You cannot detect any change in location as the car or star moves along your line of sight.

In order to measure this radial component of stellar motion, visible starlight has to be dispersed into its different colors and dark features detected within them. The wavelengths of these spectral lines are well known for a non-moving star, and the *radial velocity* component of a moving star is determined from the way its observed line wavelengths have been shortened or lengthened by its movement.

The Austrian scientist Christian Doppler first suggested such a wavelength shift for sound waves, when he also proposed a change in the colors of stars produced by their motion relative to the Earth.[12]

The change in the wavelength of radiation due to the relative radial motion between a star and its observer, along the line of sight, is now called the *Doppler shift*. If the relative motion is toward the observer, the radiation waves tighten up and shift to shorter wavelengths, and when that motion is away the waves stretch out with longer lengths (Fig. 5.3). The greater the radial velocities along the line of sight in either direction, the bigger the wavelength-shift.

The first successful measurements of the radial line-of-sight velocities of stars did not occur until the 1890s at the Potsdam Astrophysical Observatory in Germany, when Hermann Vogel and Julius Scheiner used photography with a refractor only 11-inches (0.28 meters) in diameter to obtain the long exposures needed to record their spectra. For several stars they obtained radial velocities of up to 30 kilometers per second with respect to the Sun, with an uncertainty of about a tenth that value.[13] That was nearly two centuries after Edmond Halley's discovery in 1718 of the transverse component of a star's motion.

Once much larger telescopes were constructed and the relevant instruments had been devised, the number of measurements of stellar motions increased dramatically. During the 20th century, many astronomers dedicated their lives to measuring the positions, proper motions, and radial velocities of tens of thousands of stars, resulting in extensive catalogues of these quantities.

The Sun's Motion

The observed transverse and radial velocities of stars are relative measurements. They can be due to the Sun's motion, the star's motion, or some

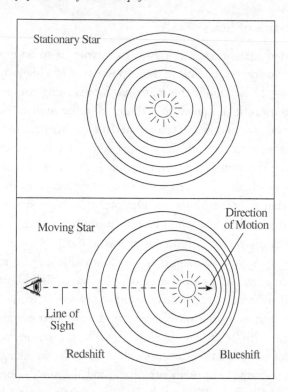

Figure 5.3 Doppler effect A stationary source of radiation (*top*) emits regularly spaced light waves that get stretched out if the source moves away from the observer (*bottom*). The size of the wavelength change from the stationary to moving condition provides a measurement of the relative speed of the source's motion along the line of sight.

combination of both of them. To separate the stellar and solar motions, the detected movements are compared to the mean of all of those observed. The mean result is attributed to the Sun's motion with respect to the nearby stars, and the motion of individual stars is inferred from their departures from this mean.[14]

The speed with which the Sun moves along its path among the stars can only be measured from the radial motions that have been extracted from the Doppler shifts of the stars' spectral lines. These results had to be obtained using spectral photography to increase the effective observing time of the dispersed starlight. The American astronomer William Wallace Campbell pioneered such measurements in the late 19th and early 20th centuries. He derived a speed of about 20 kilometers per second for the solar motion with respect to the nearby stars, which became widely accepted.[15]

The nearby stars are not moving much faster than the Sun with respect to their neighbors, but they are all revolving about ten times as fast around a common, distant and massive center. This brings us to the discovery that our Milky Way is much larger than it was once thought to be, and that all its stars are whirling about a remote location.

The Sun is Immersed within the Milky Way

On a clear moonless night, we can look up and see a hazy, faint, luminous band of light that stretches across the sky from one horizon to the other; it is known as the *Milky Way* (Fig. 5.4). According to ancient Greek myth, the goddess Hera, Queen of Heaven, spilled milk from her breasts into the sky. The Romans called the spilt milk the *Via Lactea,* or the "Milky Way." It is also called our *Galaxy*, with a capital G, derived from the Greek word *galakt-* for "milk."

Figure 5.4   The Milky Way A panoramic telescopic view of the Milky Way, the luminous concentration of bright stars and dark intervening dust clouds that extend in a band across the night sky. (Courtesy of the Lund Observatory, Sweden.)

For at least two thousand years, it was supposed that the Milky Way consists of stars, rather than misty white clouds. In the first century, the Roman poet Ovid, for example, wrote:

> "A way there is in Heaven's expanded plain,
> Which, when the skies are clear, is seen [from] below,
> And mortals, by the name Milky, know.
> The ground-work is of stars...."[16]

As soon as telescopes were invented, astronomers confirmed that the luminous parts of the Milky Way are mainly due to multitudes of stars too distant and faint to be resolved with the unaided eye. As Galileo Galilei noted in 1610:

> "The Galaxy is nothing else than a congeries of innumerable stars distributed in clusters. To whatever region of it you direct your spyglass [telescope], an immense number of stars immediately offer themselves to view. ... What is even more remarkable — the stars [celestial regions] that have been called "nebulous" by every single astronomer up to this day are swarms of small stars placed exceedingly closely together."[17]

By the 18th century, all stars were thought to be distant suns, and the flat shape of the Milky Way was attributed to the rotation of a vast collection of stars. In 1750, for example, the English astronomer Thomas Wright, born in County Durham, speculated that the visible, luminous arc of the Milky Way is a small part of either a vast ring or an enormous spherical shell of stars, centered on a Divine Presence, our God (Fig. 5.5).[18] The Sun and other stars were supposed to be in circular motion around this common Divine Center, the place from which God's infinite and eternal power emanates and directs the motion of the stars. Wright futher speculated that there might be other star systems, similar to our Milky Way, with their own divine centers.

In 1755 the German philosopher Immanuel Kant attributed the flattened shape of the Milky Way to its formation from a large, collapsing, rotating nebula all in accordance with the plan of the "Great Master Builder,"[19] much like the origin of our Solar System from a considerably smaller nebula. He improved on Wright's model by replacing his ring of stars with a spinning disk shape, and supposed that the stars in the Milky Way are all revolving about a

Sphere of Stars   Milky Way

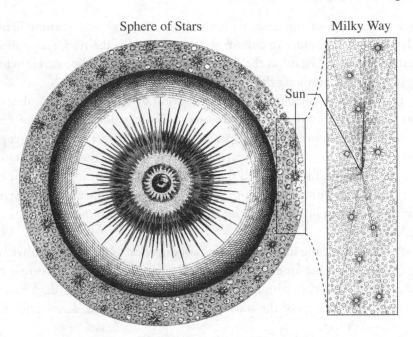

Sun

Figure 5.5   Model of the Milky Way Thomas Wright proposed that the Milky Way is composed of a large number of stars arranged in a layer, with the Sun placed at its center (*right*). It was supposed to be a small segment of a much larger spherical shell of stars, with a Divine Presence signified by a central eye (*left*). (Adapted from Thomas Wright, *An Original Theory or New Hypothesis of the Universe*, 1750, Reproduced by Science History Publications, New York 1971, page 139 plate XXV.)

very distant center, with even the nearest stars so far away from us that their motion could not be detected.

Moreover, Kant speculated, some of the fuzzy nebulae observed in the Milky Way may represent separate "island Universes," or Milky Ways, similarly composed of innumerable stars too distant to be discerned individually. Although these concepts were probably not that influential in Kant's time, since his publisher went bankrupt and only a few copies of his book reached the public, his basic ideas were revived and accepted in subsequent centuries, when astronomers realized just how prescient he was.

It was the German-born English astronomer William Herschel who provided the first observational verification of these imaginative speculations, just a few decades after Wright and Kant proposed them. He constructed the biggest telescopes of the time, with the largest mirrors and greatest light-gathering power, in order to fathom the distribution of the stars, gauge their depth in space, and establish the shape of the stellar Milky Way. As Herschel put it, he was determining the "Construction of the Heavens."

Herschel established the stellar distribution by counting the number of stars he could see in different directions. Assuming that the stars are distributed fairly uniformly in the space they occupy, and that his telescope could penetrate to the faintest stars at the boundary of the stellar system, then the number of stars in each field of view would indicate the extent of the system in that direction. In other words, the greater the number of stars seen along a given line of sight, the larger the distance to the outer edge of the Milky Way.

After counting the stars in thousands of directions, and avoiding obvious star clusters, Herschel found in 1785 that the Sun is located at the center of a flattened disk of stars with a disk diameter that is 5 times its thickness.[20] But since the distances of the stars were not known, no one knew how big the stellar disk was.

Wright, Kant, and Herschel were all correct in supposing that the Sun is one star of many, all immersed in the flattened disk of the Milky Way, but Herschel was wrong about its center. Determinations of the distances to globular star clusters were eventually used to infer the great extent of the stellar system, and to demonstrate that the Sun is not located at its center.

Harlow Shapley, the Wonder of the Whole Natural World

Harlow Shapley was a country boy, born November 2, 1885 on a farm near Nashville, Missouri. At the age of 15, he became a crime reporter for the *Daily Sun* in Chanute, Kansas, a rough oil-mining town, and went on as a police reporter for the *Times* at Joplin, Missouri, an even tougher town. After saving his money, he entered the Presbyterian Carthage College Institution with only a fifth-grade education. Harlow then completed six years of high school training in one year and a half and graduated in 1907.

Shapley then went on to Missouri University at Columbia where he intended to study at its new school of journalism. But when he got there, Shapley found that the opening of the much-advertised school of journalism had been put off for another year. As he tells it, when he looked into the university catalogue, he could not even pronounce the first entry, archeology, and so he picked the second one, astronomy, beginning this career almost by accident. A few years later, he discovered that the Thaw fellowship in astronomy was available at Princeton University, and continued his studies there; mainly for financial reasons.[21] Such unexpected turns of events have probably played important roles in most of our lives.

At Princeton, Shapley became the first graduate student of Henry Norris Russell, and embarked on a four-year investigation of eclipsing binary stars. Then in 1914, Harlow married Martha Betz, who he had met at the University of Missouri, and they traveled to his first astronomical job at the Mount Wilson Observatory in California. In the next few years at Mount Wilson he completed his legendary determinations of the distances and distributions of the globular star clusters, which displaced the Sun from the center of our stellar system and greatly enlarged its known extent.

During his years of nighttime observing from Mount Wilson, Shapley frequently spent his days observing ants. He found that trail-runner ants move back and forth along well-defined paths with speeds that increase with their temperature, and even published the result in the *Proceedings of the National Academy of Sciences*.[22] The ants, he noted, are highly civilized, altruistic, and loyal to their home, but have little hope of venturing outside their trail and escaping from their rut of uniformity. A similar fate, he thought, threatens many graduate students.

The death of Edward C. Pickering in 1919 resulted in a search for his successor as Director of the Harvard College Observatory. The ambitious Shapley had already informed Harvard authorities of his interest in the position at least a year before Pickering's demise, and upon hearing of the event he promptly wrote letters requesting support for his replacing Pickering to his Princeton mentor and to George Ellery Hale, the Director of the Mount Wilson Observatory where he then worked. Russell replied that he would not recommend Shapley to take Pickering's position, and that "you would make the mistake of your life if you tried to fill it." Hale recommended that Shapley should never attempt to take an active part in seeking the post, but the next year he supported Shapley's application for it, based on his knowledge, ability, industry and daring.[23]

In 1920, Russell turned down an offer for the position, and Shapley was then considered an important candidate.[24] After alternative choices were considered, Shapley was offered a temporary position beginning in April of 1921, and soon became Professor of Astronomy at Harvard University and Director of the Harvard College Observatory, a post he held for thirty-one years.

When remembering Shapley, Harvard Astronomy Professor Chuck Whitney wrote: "I have never seen a quicker mind, a more agile sense of humor, or a more complete absence of what usually passes for humility."[25] Harlow seems to have been gifted, ambitious, and hard working, with a tendency to avoid recognition of other astronomers, even those whose findings were directly related to his own.

In the late 1940s and early 1950s, conservative congressmen thought his outspoken liberal views and distrust of authority were dangerous and even subversive, including the infamous Senator Joseph McCarthy who listed Shapley as a Communist — which he was not.

Harlow Shapley was not a devout man in any traditional religious sense, but he had respect for religious movements. He rarely mentioned God, and does not appear to have either believed in a Deity one might pray to or to consult the *Bible* for inspiration. In his view, belief in the supernatural had to be tempered with rational thought, and living things were not in need of divine interventions. Shapley nevertheless thought the Divine might be found in Nature, writing: "It is a religious attitude to recognize the wonder of the whole natural world ... to avow reverence for all things that exit, all that is touched by cosmic evolution, and reserve the greatest reverence of all for existence itself."[26]

Shapley was much taken by the relatively recent discoveries that heavy elements are synthesized within stars. "Every baby born, every saint and sinner, and every common man and common beast breaths some of these former elements of the stars," he exclaimed. "These elements have already participated in the "snorts, sighs, bellows, shrieks, cheers, and spoken prayers of the prehistoric and historic past."[27]

The Harvard astronomer endorsed a religion that includes our new-found knowledge of the observable Universe, with its abundant mysteries that still lie beyond our grasp, and urged everyone to participate in the search for the unseen and unknown.[28]

In his later life Shapley played an important role in the founding of IRAS, or the Institute for Religion in an Age of Science, an institution that survives to this day. One of the institute's objectives is to combine a scientific understanding of the natural world with the goals and hopes of humanity expressed in religion. These purposes resonated with Shapley's proposals that traditional Christianity could be enriched and vitalized by including the discoveries of modern science and that religion might also ennoble the concepts of science.

To illustrate an overlap of science with religious concerns, Shapley quoted four pages of an address in 1951 by Pope Pius XII to the Vatican Academy of Sciences, which endorsed scientific observations of ongoing change throughout the living and non-living Cosmos.[29] But Shapley didn't include a complete account of the Pontiff's remarks, which were entitled: "The Proofs for the Existence of God in the light of Modern Natural Science."

The Pontiff was specifically referring to the expanding Universe that could be extrapolated back in time to a beginning. If there was a beginning, he argued, then there had to be a Creator God, as First Cause. This would explain the origin of the outward motion of spiral nebulae [galaxies] and all subsequent transformations in the ever-changing Universe.

Shapley wrote several essays in the 1960s that touched on the interface of astronomy and religion. At this time in his life, when he was more than 75 years old, Shapley also participated in considerations of religion in the scientific age, edited the book *Science Ponders Religion*, and proudly received a Doctor of Divinity from the Meadville-Lombard Theological School affiliated with the University of Chicago.

As Shapley's Princeton mentor Henry Norris Russell had also mentioned, a transcendent God is the only reason why there is any Universe.[30] Science, Russell noticed, answers only the question of *how* things come to pass, but not *why* things are so.

Enlarging and Re-Centering the Milky Way

Shapley's greatest contribution to astronomy was the discovery that globular star clusters could be used to look outside the flattened disk of the Milky Way to establish its dimensions and center. Each of these dense clusters contains hundreds of thousands of stars held together and bound into a spherical shape by their mutual gravitation (Fig. 5.6). Observing them was analogous to flying in an airplane to look down and determine the extent of a city,

Figure 5.6   Star cluster Several hundred thousand stars swarm around the center of the globular star cluster NGC 6934, which is estimated to be about 10 billion years old. (A *Hubble Space Telescope* image courtesy of NASA/ESA.)

which cannot be done from inside where nearby buildings hide the distant parts of the city from view.

Harlow's doctoral thesis at Princeton University, published in 1913, involved nearly 10,000 observations of eclipsing binary stars, which enabled him to determine the orbital parameters of 90 of them and to increase the number of known orbits by about ten times.[31] When he arrived at the Mount Wilson Observatory in 1914, Shapley turned his attention to another type of variable star known as the Cepheids, after their prototype Delta Cephei. This class of variable stars changes periodically between a bright state and a dimmer one and back to a bright condition again. These stars are additionally distinguished by an exceptional luminosity, high temperature, and large mass and size. The Cepheid stars are so big, Shapley showed, that if they were eclipsing binary stars the two hypothetical stars would have to be inside each other. As an alternative, Shapley suggested that the Cepheid variations might arise from pulsations of isolated individual stars.[32] As the outer atmosphere of a Cepheid contracts and expands with a regular beat, it acts like a valve that periodically absorbs and releases the outward flow of energy from the star's central region, leading to its periodic light variation.

While Shapley was investigating eclipsing binary stars back at Princeton, Henrietta Leavitt, a researcher at Harvard College Observatory, discovered a novel way of determining the distances of these luminous Cepheid variable stars. After years of scrutiny, she had determined the period of light variation for 25 variable stars in the Small Magellanic Cloud, which ranged from 1.2 to 127 days, and showed that this variation period increases with the star's observed brightness. Since all of these Cepheids were located within the remote Small Magellanic Cloud, she could therefore modestly state that: "Since the variables are probably at nearly the same distance from the Earth, their periods are apparently associated with their actual emission of light [or luminosity]."[33] The more luminous a Cepheid variable star is, the more slowly it varies and the longer its variation period.

When he arrived at Mount Wilson in 1914, Shapley began a study of the Cepheid variable stars in globular star clusters, and found that for a given star cluster they displayed a period-luminosity relation similar to the one that Leavitt had found for the Cepheids in the Small Magellanic Cloud, and concluded that their periodicities and apparent brightness could be used to establish the relative distances of the star clusters. It remained for Shapley to determine the mean luminosity of nearby Cepheids of reliably known distances, and thereby infer the distances of remote Cepheids in globular clusters.[34] This mean luminosity is roughly 10,000 times the luminosity of the Sun.

Since the Cepheids are so highly luminous, they can be detected at relatively large distances when compared with fainter stars. Moreover, the mean luminosity of each star is related to its pulsation period, which provides a way of estimating its distance from its observed brightness and established luminosity.

In these ways, Shapley discovered the enormous distances of the globular star clusters. In October 1917 he wrote to his Princeton mentor Henry Norris Russell of this "peculiar Universe" in which the nearest globular clusters were about 20,000 light years away and the furthest ones something like ten times that amount.[35] In the following year he reported that the globular star clusters are typically about 50,000 light years away from the Sun and thus outside the bounds of Kapteyn's Milky Way Universe.[36] [A light year is the distance light travels in one year, which is about 10 million billion, or 10^{16}, meters.]

The following year, Shapley showed that the globular star clusters form a vast, roughly spherical system enveloping the plane of the Milky Way on both

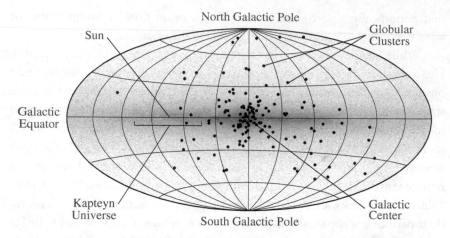

Figure 5.7   Edge-on view of the Milky Way The globular star clusters are distributed in a roughly spherical system that envelops the Milky Way and is centered at about 27,700 light-years away from the Sun. The disk and central bulge are shown edge-on in an infrared image that penetrates the interstellar dust that limits an astronomer's view at optically visible wavelengths to a much smaller Kapteyn Universe, centered on the Sun.

sides and centered about 65,000 light years away from the Sun in the direction of Sagittarius (Fig. 5.7).[37] The diameter of the system is, he reported, some 300,000 light years in the plane of the Milky Way.

At the time, the accepted maximum extent along the plane of the Milky Way corresponded to a diameter of perhaps one hundredth of Shapley's value. In 1914, A. S. Eddington had estimated a total extent of perhaps 3,000 light years, and on hearing of Shapley's awesome feat he wrote him exclaiming that:

> "I think it is not too much to say that this marks an epoch in the history of astronomy, when the boundary of our knowledge of the Universe is rolled back to a hundred times its former limit."[38]

By looking up, out and beyond the known stellar system, Shapley had increased the total known extent of the Milky Way at least tenfold, and its volume more than a thousand times. It was a remarkable finding applauded by some of the most eminent astronomers of the time. In addition, Shapley removed the center of the known stellar Universe from at or near the Sun, and placed the Solar System far off center in the peripheral outer fringes of the Milky Way.

Shapley's enlarged and re-centered Milky Way was not initially accepted by some established astronomers of the time. In the early 20th century, the Dutch astronomer Jacobus C. Kapteyn and his colleagues had exploited long photographic exposures to extend star counts to the faintest possible limits, and used measurements of the distances of some of the nearest stars to provide a scale to their model. In 1922 Kapteyn concluded that the stars in the Milky Way reside in a Sun-centered system and are confined within the plane of the Milky Way out to a maximum distance of 30,000 light-years from the Sun.[39]

As it turned out, astronomers could only discern the nearby parts of the Milky Way when looking directly into it. The more distant invisible parts lie behind an opaque veil of interstellar dust that absorbs the light of distant stars. So looking deep within the Milky Way is something like viewing distant objects on a foggy day. At a certain distance, the total amount of fog you are looking through mounts up to an impenetrable barrier. This meant that Shapley was right about the larger extent and remote center of the Milky Way, and that Kapteyn just couldn't see that far.

Where is the center of the Milky Way, and how far away is it? According to Harlow Shapley, this invisible center has to be located at the heart of the globular cluster system some tens of thousands of light-years away in the direction of the constellation Sagittarius. Such a remote location would be completely obscured by an intervening curtain of interstellar dust, and therefore hidden from direct observations with visible-light [optical] telescopes.

The discovery of the unseen center had to await the development of new technology at invisible radio and infrared wavelengths, initially for military purposes. Radiation at these longer wavelengths, beyond those normally visible to the eye, is able to penetrate and see through the interstellar dust. In much the same way, radio waves pass through storm clouds to reach your car radio or cell phone, even when it is raining or snowing. Observations at these invisible wavelengths have enabled astronomers to conclude that the Sun and nearby gas and stars are revolving about the center of the Milky Way at a speed of 220 kilometers per second, with an orbital period of about 240 million years at their distance from the center of 27,700 light-years.

The most intense radio emission is also coming from the direction of the constellation Sagittarius.[40] This compact radio source coincides with a distant collection of infrared-emitting stars,[41] and is no larger than our Solar System.[42, 43] It is most likely energized by a massive black hole that is a colossal 4 million times the mass of the Sun.[44]

As we shall next see, hundreds of billions of whirling spiral nebulae are moving away from the Milky Way and each other at speeds that increase with their distance. This discovery involved an extension of Shapley's observations of the Cepheid variable stars in globular star clusters to Cepheids found in spiral nebulae, and disagreed with Shapley's belief that they had to be embedded within the Milky Way.[45]

6. The Universe is Expanding and Breaking Away

"Before their eyes in sudden view appear
The secrets of the hoary Deep — a dark
Illimitable ocean, without bound,
Without dimension; where length, breadth, and heighth,
And time, and place, are lost."

John Milton (1667)[1]

Spiral Nebulae, a Tale of Larger Telescopes

In the 17th and 18th centuries, telescopes revealed two types of cosmic objects that could not be distinguished with the unaided eye. They were the bright star clusters and the pale, cloudy nebulae. Charles Messier listed the 103 most prominent in his catalogue of 1781,[2] and today they remain designated by the letter "M" followed by the number in his catalogue.

When telescopes were constructed with larger mirrors than ever before, they were naturally used to scrutinize the nebulae in Messier's list in detail, and to see if they could be resolved into component stars or remained of misty disposition under close scrutiny. As an example, the English astronomer, William Herschel, used his most productive telescope, with a metal mirror of 18.7-inches (0.41 meters) across, to show in 1784 that many, but not all, of Messier's round nebulae could be resolved into globular star clusters. His mirror was about five times bigger than that of his contemporary Messier.

Sixty years later, William Parsons, third Earl of Rosse, constructed an even bigger telescope at the family's Birr Castle in Ireland (Fig. 6.1). The

Figure 6.1   Leviathan of Parsonstown William Parsons, the third Earl of Rosse, surveyed the Heavens with this telescope from the grounds of Birr Castle, his family's ancestral estate in Northern Ireland. The four-ton, 1.8-meter (72-inch) diameter speculum mirror, made of an alloy of copper and tin, was the largest in the world from 1845 to 1977, when it was surpassed by the 2.5-meter (100-inch) Hooker telescope at the Mount Wilson Observatory near Pasadena, California. (Courtesy of the Trustees of the Science Museum, London.)

metallic mirror of the *Leviathan of Parsonstown*, as it was known, had a diameter of 6 feet, or 1.8 meters, and four times the diameter and sixteen times the light collecting area of Herschel's instrument.

With the added magnification of his gargantuan reflector, Rosse discovered, in the spring of 1845, an entirely new kind of cosmic object, the spiral nebulae. He showed, for example, that the fifty-first nebula in Messier's catalogue, denoted M 51, has the spiral shape of an immense whirlpool, which Rosse attributed to its rotation. Since photography had not yet been introduced to astronomy, Rosse used fine drawings to display the structure of M 51 (Fig. 6.2), and fourteen other spirals.[3]

Figure 6.2 **Spiral nebula** The curved shape of the Whirlpool Galaxy, M 51, is illustrated in this drawing made by Lord Rosse when using his 1.8-meter (72-inch) diameter telescope in the spring of 1845; he subsequently found at least a dozen other nebulas with a spiral shape. [Reproduced from The Earl of Rosse, "Observations of the Nebulae," *Philosophical Transactions of the Royal Society*, pages 110–124, plate 35 (1850).]

In the closing years of the 19th century, long exposures of photographic plates transformed the way the faint spiral nebulae were discovered and studied (Figs. 6.3, 6.4, 6.5). By 1900, the American astronomer James E. Keeler had conservatively estimated that 120,000 spiral nebulae were within the reach of his photographic telescope at the Lick Observatory in California, even though it was half the size of Lord Rosse's mirror.[4] At about the same time, other astronomers had begun to speculate that the Milky Way also has a spiral structure.[5]

Most of the thousands of newly discovered spiral nebulae were very faint and narrow in angular extent, which suggested that they are far away. But no one knew for certain whether the spirals are enormously large and exceedingly distant collections of innumerable stars, or small, nearby objects, little spinning wheels of

Figure 6.3   Edge-on spiral The galaxy NGC 4565 is portrayed in this image taken from the 2.1-meter (82.7-inch) diameter telescope at the Kitt Peak National Observatory. (Courtesy of Bruce Hugo and Leslie Gaul/Adam Block/NOAO/AURA/NSF.)

gas that swirl in the interstellar spaces of the Milky Way. Their composition and structure was a matter of speculation until direct measurements of their exceptionally high speeds and enormous distances were obtained. These findings indicated that the spiral nebulae are huge stellar systems separated by vast stretches of empty space and that they are all participating in the expansion of the Universe.

A Quiet, Modest Man and a Vain, Ambitious One

Two Americans, V. M. Slipher and Edwin P. Hubble, jointly contributed to the discovery of the expanding Universe, although neither one of them ever claimed credit for it. One of them, Slipher, was a modest and private man working at an observatory that was mainly dedicated to searching for signs of life on Mars. The other astronomer, the pipe smoking, fly-fishing Hubble, was vain and egocentric, with access to the most powerful telescopes of his time. As far as the general public was concerned, Slipher remained largely

Figure 6.4 Face-on spiral The Pinwheel galaxy M 101 has spiral arms that are about 170,000 light-years across. This portrayal is a composite of *Hubble Space Telescope* images superimposed on ground-based images taken at the 142-inch (3.6-meter) diameter Canada-France-Hawaii Telescope in Hawaii and at the 35-inch (0.9-meter) diameter telescope at the Kitt Peak National Observatory. (Courtesy of NASA/ESA/STSci/NOAO/AURA/NSF, K. Kuntz (JHU), F. Bresolin (U. Hawaii), J. Trauger (JPL), J. Mould (NOAO), Y.-H. Chu (U. Illinois, Urbana), J.-C. Cuillandre (CFHT), and G. Jacoby, B. Bohannan, and M. Hanna (NOAO).

unknown throughout his life, and became one of the unsung heroes of American astronomy. In contrast, Hubble was one of the most renowned astronomers of the 20th century.

Vesto Melvin Slipher, who was almost always referred to as "V. M.," was born on a farm in Mulberry, Indiana on November 11, 1875, the second of nine children that grew to maturity. He graduated from high school in Frankfort, Indiana, taught briefly at a nearby country school, and at the age of 21 entered the Indiana University in Bloomington, where he studied mechanics and astronomy.

In 1901 Slipher began a fifty-three year career as an astronomer at the Lowell Observatory in Flagstaff, Arizona. He also acquired extensive property

Figure 6.5   Endless galaxies A remote cluster of galaxies, designated CL 0939 + 4713 for its coordinates on the sky, as it looked about 10 billion years ago when the light we see was emitted and the Universe was two-thirds of its present age.   (A *Hubble Space Telescope* image courtesy NASA, STScI, and Alan Dressler, Carnegie Institution.)

in and around Flagstaff, eventually bought a number of ranches, owned and operated a retail furniture store, and managed many rental properties.[6] V. M.'s younger brother, Earl Carl, became a staff astronomer at the same observatory in 1905, and eventually served as Arizona State Representative, Mayor of Flagstaff, and Arizona State Senator.

Percival Lowell, a wealthy Bostonian, funded the construction of the Lowell Observatory in order to observe Mars in the clear, cool air of the San Francisco Mountains near Flagstaff. He believed that intelligent Martians had constructed canals on the planet to transport water from its melting polar caps to parched equatorial deserts, and wrote popular books about them. At the time, the spiral nebulae were thought to resemble our Solar System in its early formative stages, and Lowell instructed V. M. to examine the spiral nebulae to find out more about the beginning of our own planetary system.

When he observed the spectral lines of the brightest spiral nebulae in 1914, Slipher unexpectedly discovered that most of them are moving away from us at velocities well in excess of those of any other known cosmic object. Although these results did not become widely known, they did have a noticeable impact on several astronomers who corresponded with V. M. and traveled to Lowell Observatory to find out more about his velocity measurements.

After Slipher finished his observations of bright spiral nebulae, the more flamboyant Edwin Powell Hubble determined their remote distances and received most of the public attention.

Hubble was born on November 20, 1889 in Marshfield, Missouri and grew up there and in Wheaton, Illinois, a suburb of Chicago, with a comfortable life as the son of a successful insurance executive. As a teenager, he earned spending money delivering morning papers. He also had a stern father who gave him "good lickings," which he later said "did him a lot of good."[7]

Edwin was a strong, gifted, and confident high-school athlete, and a smart, charismatic, handsome, polished and self-assured young man. His biographer, Gale E. Christianson, has also portrayed him as ambitious, insensitive and shallow, a social climber and a liar.[8]

Hubble attended the University of Chicago, receiving his undergraduate degree in mathematics and astronomy in 1910. In the same year, he was elected as a Rhodes scholar despite a B– undergraduate grade average, at the recommendation of the eminent physicist Robert A. Millikan, who Edwin had worked for as a laboratory assistant. [Hubble attended Queens College at Oxford University in England, where he was trained in English law and also studied literature and Spanish.]

In 1913 Edwin joined his family in Louisville, Kentucky, where he became a high school teacher for a year, and at the age of 24 entered graduate school at the University of Chicago to study astronomy at its nearby Yerkes Observatory; his subsequent doctoral thesis in 1917 concerned photographic investigations of faint nebulae.[9] Edwin then joined the American Army, and with the rank of Major was sent to France in September 1918 a few months before the end of the First World War (1914–1918). Hubble enjoyed the plain hard living, the simple food, the discipline, and the adventure of army life.[10]

When the war was over, he returned to the United States, and in 1919, at the age of 30, began a position as staff astronomer at the Mount Wilson Observatory near Pasadena, California. He was one of the first to use the unprecedented light-collecting power of the observatory's 100-inch (2.5-meter) reflector to explore the realm of the faint spiral nebulae, and pioneered determinations of the distances to the brightest ones.

Hubble met Grace Burke, the daughter of a wealthy Los Angeles banker, when she was staying at the Kapteyn cottage on Mount Wilson with the wife of another astronomer. At the time, Grace was married to the geologist Earl Lieb, also from an established California family — his father later became president of Stanford University. When Grace's husband died, from inhaling carbon monoxide while exploring a coal mine, Grace and Edwin began a romance that culminated in their marriage in 1924.

By the following year, Hubble had found that the bright spiral nebulae are too distant to be located in the Milky Way, which indicated that the observable Universe is much larger than many astronomers thought it was at the time. Five years later, he showed that the greater the distance of a spiral nebula, the faster it is moving away from us. This meant that the known Universe is getting bigger as time goes on, and also established its great but finite age.

Edwin's observations made him famous, and he and his wife Grace relished the notable friendships that resulted from his notoriety. They socialized with the likes of Charlie Chaplin, Frank Capra, Paulette Goddard, Harpo Marx, and William Randolph Hearst, and were constant companions of Aldous and Maria Huxley who shared their disdain for the lower class. Hubble appeared on the cover of the February 9, 1948 issue of *Time* magazine, was photographed with Hollywood stars, and well after his death NASA named its *Hubble Space Telescope* after him.

Despite his fame, Hubble held whatever beliefs he had very close to his chest. When asked about them, he replied that: "The whole thing is so much bigger than I am, and I can't understand it."[11] On another occasion, Hubble wondered about the relationship of his observations to an understanding of God, and stated that: "We do not know why we are born into the world, but we can try to find out what sort of a world it is — at least in its physical aspects."[12] To his surprise, Hubble's name and observations were even mentioned by Pope Pius XII, in a 1951 address entitled "Proofs for the Existence of God in the light of Modern Natural Science."[13]

In his later years, Hubble noted that the scientific realm is restricted to just one aspect of the Universe, which deals with probable knowledge of the observed world of Nature. This knowledge comes from observation and experiment, and can be easily communicated and tested. There is another world, he supposed, that science cannot enter, one of eternal, ultimate truth. "Sometimes, through the strangely compelling experience of mystical insight," Hubble wrote, "a man knows without the shadow of doubt, that he has been in touch with a reality that lies beyond mere phenomena."[14] This ecstasy, this sense of wonder, he declared, is a private revelation that can only be understood by those who experience it.

Hubble's assistant Allan Sandage related this awe and wonder to a religious sentiment, commentating that: "There has to be some organizing principle. God to me is a mystery but is the explanation for the miracle of existence — why there is something rather than nothing."[15]

Not much was said, as far as this author can discover, about Slipher's beliefs. Like Hubble, he was mainly concerned with observations of the natural world, so lets get on with his unexpected discovery of the high velocities of spiral nebulae.

The Spiral Nebulae are Moving Very Fast

Vesto M. Slipher, known as V. M., made the unanticipated discovery that spiral nebulae are moving at exceptionally high speeds. Many astronomers then thought that the bright centers of spiral nebulae were newborn stars, and that the surrounding spiral arms were nascent planetary systems, which whirled and rotated around the central star just as the Earth revolves about the Sun. So Slipher set out to observe their rotation, but that is not what happened! He discovered that they were nearly all moving away from the Earth and each other at velocities of up to 1,100 kilometers per second, and much faster than any star.

Slipher reported his extraordinary findings for 15 spiral nebulae in August 1914 at the 17th annual meeting of the American Astronomical Society. With characteristic modesty, he stated that: "In the great majority of cases the [spiral] nebula is receding ... the striking preponderances of the positive sign [outward velocity] indicates a general fleeing from us or the Milky Way."[16] His astonishing discovery received a standing ovation from the audience.

For the next decade, Slipher worked almost alone in his pioneering measurements of the high-speed motions of spiral nebulae, and it wasn't easy.

Heroic exposure times and precise hand-guiding of his telescope were required for periods of 20, 40 and even 80 hours to obtain the elusive spectra of the faint spirals, and to infer radial velocities from them. Few such observations had therefore even been attempted, and it explains why almost no one else tried the measurements for many years.

Although the brilliant Andromeda nebula, denoted M 31, is moving toward us, most of the brightest spiral nebuae that were within the range of Slipher's small 24-inch (0.6-meter) refractor are moving away. By 1917 he had discovered that at least 20 of them are in outward motion.[17]

After six more years he had discovered that 41 out of 45 spiral nebulae are fleeing from us and each other. By this time, he had largely completed observations of the brighter spiral nebulae whose spectra could be measured using his small telescope; so Slipher moved on to other things, including spectroscopic observations of the aurora, comets, planets, the night sky, and stars, as well as the search for a distant Planet X beyond Neptune.

It has been suggested that Slipher's discoveries did not become well known because he worked in the remote, scientific backwaters of the Lowell Observatory whose primary activity involved searching for life on Mars, rather than the mainstream astronomy that was being carried out at other places such as the Mount Wilson Observatory in California. But that is not the case. V. M. traveled to Mount Wilson, in 1918 and 1921, and many prominent astronomers visited him to discuss his measurements of spiral nebulae between 1922 and 1925, including Knut Lundmark of the Observatorium at Uppsala, Sweden, A. S. (Arthur Stanley) Eddington from the University of Cambridge, Georges Lemaître of the University of Louvain, Belgium, and Edwin Hubble from Mount Wilson. Slipher also mailed his radial velocity measurements to these visitors, and others such as Gustaf Strömberg in Sweden and Harlow Shapley in the United States.

Moreover, Henry Norris Russell, one of the most influential American astronomers of the first half of the 20th century, was exceptionally familiar with Slipher's discovery. Russell returned frequently to the Lowell Observatory for refreshing summer visits with his family in the 1920s and 1930s. They would take camping trips with V. M. and his wife to places like the Canyon de Chelley, the Grand Canyon, and the Painted Desert. Young astronomers who lived on the Observatory's Mars Hill dated Russell's two daughters, Lucy and Margaret, and Margaret, the youngest, married one of them — Frank Edmondson in 1934.

Slipher's unexpected discovery of the high-speeds of spiral nebulae was therefore well known by contemporary astronomers, but their implications remained controversial. Russell could not believe that almost all of them were moving away from us, and neither could A. S. Eddington, who proposed that there must be a non-moving solution related to the curvature of space-time.[18]

The spiral nebulae might or might not be fleeing rapidly away from the Milky Way, but at the time of Slipher's discovery no one yet knew how big they were or how far away.

The Spiral Nebulae are Very Distant

If the spiral nebulae are moving at high velocity, they could not be gravitationally confined within the Milky Way, at least for very long. Already in 1914, for example, Ejnar Hertzsprung of the Astrophysikalisches Observatorium in Potsdam, Germany had written to Slipher to congratulate him on his "beautiful discovery of the great radial velocity of some spiral nebulae. It seems to me," he wrote, "that with this discovery the great question, if the spirals belong to the system of the Milky Way or not, is answered with great certainty to the end, that they do not."[19] Nevertheless, some astronomers thought that the spiral nebulae ought to be confined within the Milky Way.

The issue was presented during the now-famous Shapley-Curtis debate over "The Scale of the Universe" during a meeting of the National Academy of Sciences held in April 1920,[20] but no one could settle the controversy for certain because the distances to the spirals had not been directly measured. Subsequently, Edwin Hubble was able to use the new 100-inch (2.5-meter) reflector on Mount Wilson to detect Cepheid variable stars in bright nearby spiral nebulae, which permitted measurements of their enormous distances.

In late 1923 he decided to look for novae in the Andromeda nebula, M 31, and accidently discovered that a suspected nova varied in brightness like a Cepheid variable star with a long period of about a month. But this important discovery was not immediately accepted. On receiving Hubble's letter reporting the unexpected finding, in February 1924, Harlow Shapley replied that it was "the most entertaining piece of literature I have seen for a long time."[21] To Shapley, then the Director of the Harvard College Observatory,

Figure 6.6 Andromeda The nearest spiral galaxy, the Andromeda Nebula, M 31, as photographed using the 100-inch (2.54-meter) diameter reflector of the Mount Wilson Observatory in California. Two smaller galaxies are also shown in this image. (Courtesy of the Mount Wilson Observatory.)

Hubble's finding was just a big joke, apparently because Shapley was convinced that the spiral nebulae did not contain stars.

Undaunted, Hubble began a detailed hunt for Cepheids in M 31 (Fig. 6.6), as well as in the great spiral in Triangulum, M 33. These two had the largest angular extents of the known spiral nebulae, and were presumably the closest of the many thousands of spiral nebulae that had by then been photographed. By exploiting the light gathering power of the great 100-inch reflector, Hubble could detect the light variation of individual Cepheid stars in these two close spiral nebulae, which had never been done before.

Night after night, he used the enormous reflector to photograph the two spiral nebulae. By comparing hundreds of photographs, he found

Cepheids that periodically brighten and dim like clockwork. Within each spiral, the stars of longer variation period were the brighter ones, which was consistent with the increase of luminosity with period that Harlow Shapley had found for Cepheids in globular star clusters over a range of shorter periods of days.

By the end of 1924, Hubble was able to measure the variation periods of enough Cepheids to infer the distances of M 31 and M 33 using Shapley's period-luminosity relations for Cepheids in globular star clusters. Once the Cepheid luminosity was established from its variation period, that luminosity could be combined with the observed brightness to determine the distance. Hubble found that both M 31 and M 33 were at a distance of nearly a million, or 1,000,000, light-years. Both spirals had to reside outside the Milky Way!

This was an astounding discovery, and Hubble knew it. So he had these findings published in an article in *The New York Times*, on November 23, 1924, and then read *in absentia* by the Princeton astrophysicist Henry Norris Russell on New Year's Day, 1925, at the thirty-third annual meeting of the American Astronomical Society in Washington, D.C.[22] The historic paper, entitled "Cepheids in Spiral Nebulae," caused an overwhelming sensation.

Hubble had broken through the ancient stellar Heavens and moved the outer boundary of the observable Universe far beyond it. The Cosmos had to be much larger than anyone had previously demonstrated, and it transformed astronomers' view of the Universe. The Milky Way was relegated to just one of a multitude of spiral nebulae separated from each other by immense regions of apparently empty space.

Although Hubble failed to acknowledge it, the Estonian astronomer Ernst Öpik had inferred a large distance for M 31 three years before Hubble. While at the Armagh Observatory in Northern Ireland, Öpik used the rotation velocity of M 31 to estimate its mass at about 2 billion Suns. By assuming that its mass to luminosity ratio is comparable to that of the stars in the Milky Way, he obtained the luminosity of M 31 and combined that with its observed brightness to determine in 1922 a distance of about 1.5 million light-years.[23]

Hubble did not even mention or reference Öpik's previous, related work, but Hubble would soon make an even more dramatic discovery. The outward velocities of spiral nebulae increase with their distances, which suggested that the Universe is blowing itself apart.

Discovery of Cosmic Expansion

By 1929 Hubble had used the unparalleled 100-inch reflector of the Mount Wilson Observatory to measure the distances of 24 spirals, which were all incredibly far away and even more distant than Andromeda. When he compared these distances to their radial velocities, mainly provided by V. M. Slipher, he found that the more distant a spiral nebula is, the faster it is rushing away from us, at least to a velocity of about 1,000 kilometers per second (Fig. 6.7).[24] Not only was the observable Universe far bigger than had previously been thought, it was also expanding and carrying the spiral nebulae outward in all directions, with the fastest ones having moved the greatest distance.

A velocity-distance relation had been anticipated in 1922 by the German astronomer Carl Wirtz,[25] and in 1925 the Swedish astronomer Knut Lundmark confirmed that the more distant the spiral nebula, the faster it was receding.[26] Hubble's unique contribution was the measurement of accurate distances using Cepheid variable stars, rather than just estimating the distances from the observed brightness of the spiral nebulae.

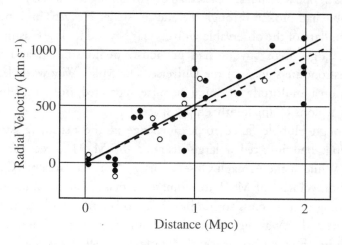

Figure 6.7   Hubble's diagram A plot of the radial velocity of nearby extragalactic nebulae, or galaxies, as a function of their distance, published by Edwin Hubble in 1929. Here the velocity is in units of kilometers per second, abbreviated km s⁻¹, and the distance is in units of millions of parsecs, or Mpc, where 1 Mpc is equivalent to 3.26 million light-years. [Adapted from Edwin P. Hubble, "A Relation between Distance and Radial Velocity among Extra-Galactic Nebulae", *Proceedings of the National Academy of Sciences* **15**, 168–173 (1929).]

With outlandish arrogance, Hubble failed to acknowledge, in his renowned 1929 paper, any previous observational evidence for a velocity-distance relation. He did not mention the published results of either Wirtz or Lundmark. Incredibly, he did not even acknowledge Slipher's velocity measurements or cite them with a reference! Hubble also used Harlow Shapley's methods of inferring cosmic distances from Cepheid variable stars.

Both Slipher's and Shapley's results had been well known by the astronomical community for more than a decade, and the failure to mention their work confirmed that Hubble was not an overly modest or generous person. His oversight was eventually corrected with acknowledgment of Slipher's pioneer work,[27,28] but Shapley subsequently noted that Hubble never did acknowledge his priority.[29]

The previous investigation that Hubble did cite was a theoretical speculation by the Dutch astronomer Willem de Sitter back in 1917. He had found that one consequence of Einstein's *General Theory of Relativity* could be an imitation motion in which more distant clocks would run more slowly; hence distant atoms, vibrating like slow clocks, would show a redshift increasing with the square of the distance.[30] In his pioneering 1929 publication, Hubble supposed that the velocity-distance relation, when corrected for solar motion, might represent this de Sitter effect, and in a letter to de Sitter, written in 1930, Hubble stated: "The velocity-distance relation among nebulae has been in the air for years — you, I believe, were the first to mention it."[31]

In 1929 Hubble was probably unaware of another relativistic derivation of the velocity-distance relation by the Belgian astronomer and diocesan priest, Abbé Georges Lemaître. It was published two years before in an obscure Belgian journal using observational results given to Lemaître when he visited both Slipher and Hubble. It is a little harder to excuse Hubble's neglect of the California Institute of Technology physicist H. D. Robertson's 1928 theoretical prediction of the relation using the available data.[32] Their offices were within walking distance of each other.

Nevertheless, most astronomers of the time were not inspired by relativity theory and theoreticians would argue for decades about which relativistic model might correspond with reality.[33] Moreover, de Sitter's interpretation without an expansion was simply wrong, and Hubble's mention of it was the last time he left the observable world to entertain theoretical speculations about it.

Astronomers now explain the velocity-distance relation in terms of an expanding Universe, in which the galaxies are all rushing away from us, dispersing and moving apart and occupying an ever-increasing volume. Yet, Hubble never did interpret his observations this way, and other eminent astronomers also wondered about such an explanation. In 1929, the renowned American astronomer Henry Norris Russell found that: "The notion that all the galaxies were originally close together is philosophically rather unsatisfactory."[34]

But Hubble knew he was onto something. With great foresight he initiated a program with Milton L. Humason using the 100-inch reflector to extend the velocity-distance relation to velocities as large as 20,000 kilometers per second and to distances as large as 100 million light-years.[35]

Milton Humason has a colorful life story. He dropped out of school at age 14, and never received a formal education after the eighth grade. He became a tobacco-chewing gambler and reputed "ladies man," and began his career at Mount Wilson as a mule driver taking construction material and equipment up the mountain. In 1917, he became janitor at the observatory, and volunteered to be a night observing assistant. In succeeding years he progressed from assistant to observer to a self-trained staff astronomer more skilled at observing than Hubble when it came to obtaining the spectra of faint nebulae.

The unimaginably high velocities indicated that the spiral nebulae, which are now known as spiral galaxies, participate in a uniform flow that gathers speed with distance. Although individual galaxies might dart here and there, even collide if near enough to each other, these localized motions are limited in velocity by gravitational interactions, and they are relatively slow when compared to the recession velocities of remote galaxies. In other words, the galaxies, or extra-galactic nebulae, have to take part in an expanding Universe, for any other plausible explanation only applies to speeds much lower than those of the most distant galaxies. The Milky Way became just one Galaxy among many, with a capital G to show it is ours.

Both Hubble and Humason were nevertheless cautious about interpreting their results and hesitant to attach any significance to them beyond the observations themselves. "The interpretation of the redshifts as velocities of recession is controversial," wrote Humason in 1931, "for the present we prefer to speak of these velocities as *apparent*."[36] And in their collaborative paper of the same year, these two astronomers described "the 'apparent velocity-displacements' without venturing on the interpretation and its cosmologic significance."[37] Hubble declined to interpret the velocity-distance relation in terms of an expanding Universe throughout his life.

He thought that cosmological models were a forced interpretation of the observational results, and as far as he was concerned theoretical cosmology consisted to a large extent of irrelevant, unverified speculations. Even as late as May 1953, shortly before his death by heart attack, Hubble was reluctant to accept the redshift as a literal expansion and referred to his famous relation as the *law of the red-shifts*, which should be formulated as an empirical relationship between observed facts.[38]

[By this time the term redshift, denoted by the lower case letter z, had entered the description; for low redshifts it is the ratio of the velocity, V, of the object to the velocity of light, c. The renowned Hubble's law states that the velocity V of a galaxy at distance D is given by $V = H_0 \times D$, where the symbol H_0 is known as the Hubble constant. It is a fundamental measure of the Universe with the same value for any galaxy.]

Hundreds of billions of galaxies have now been located with telescopic eyes, and there isn't any end in sight. They are as numerous as snowflakes in a storm or grains of sand at the seashore, and each galaxy is composed of billions of stars. And as far as we know, the Universe has no perceptible outer boundary. Moreover, the entire Universe looks practically empty, for immense regions of apparently vacuous space separate most of the galaxies from each other.

The overwhelming immensity of space had been imagined in the 17th century, by the French philosopher Blaise Pascal who wrote: "The whole visible world is only an imperceptible dot in nature's ample bosom…. Nature is an infinite sphere whose center is everywhere and circumference is nowhere…The eternal silence of these infinite spaces frightens me."[39]

Pascal was right about the dark immensity of space, its overwhelming largeness, which has been confirmed by modern astronomical surveillance of the enormously distant horizons of the observable Universe.

Children and adults can still find the darkness scary. That could be why they tell ghost stories around campfires at night. Many astronomers nevertheless revel in the dark quiet of the night and the splendor it brings into view. Their understanding of the fullness of the dark places, and the formation of stars within it, has made the spacious Universe more inclusive and less threatening.

When did the Expansion Begin?

By reversing the expansion of the observable Universe, you can find out when the galaxies started on their outward journey. Assuming a constant speed, that expansion age is obtained by dividing the distance of any remote extra-galactic nebula, or galaxy, by its velocity. It is known as the *Hubble time*, the reciprocal of the Hubble constant.

If the expansion has continued at a steady pace with the rate determined by Hubble's measurements, the expansion began about 1.8 billion, or 1.8×10^9, years ago. But that is less than the age of the crust of the Earth, accurately dated in 1956 at 4.6 billion years from the radioactive decay of its oldest rocks.[40] How could the Earth be older than the expanding Universe?

Well, it isn't! In 1952, a quarter of a century after Edwin Hubble had set up the extragalactic distance scale, Walter Baade announced that he had corrected the distance scale, owing to two different kinds of pulsating stars, and this meant that a steady expansion began about 3.6 billion years ago.[41,42] That still wasn't long enough ago, but Hubble's assistant and protégé, Allan Sandage, eventually corrected the identification of the brightest stars in Andromeda, and in 1958 announced that the Hubble time is close to 13 billion years, and significantly older than the Earth.[43] [When Harvard's Department of Astronomy invited Sandage to talk about his measurements of the Hubble constant, which was being estimated in other ways by Harvard graduate students and faculty, Sandage wrote back saying that his mother had taught him not to talk to the village idiot.[44]]

By the turn of the 20th century, the dust had settled and most astronomers now accept a Hubble time of 13.7 billion years. Wendy Freedman and her colleagues at the Carnegie Observatories in Pasadena, for example, used the *Hubble Space Telescope* to scrutinize Cepheid variable stars in nearby galaxies and to obtain in 2001 an expansion age of about 15 billion years (Fig. 6.8).[45]

The Universe is therefore considerably older than the oldest terrestrial rocks, and we can all be reassured that the expanding Universe had a definite

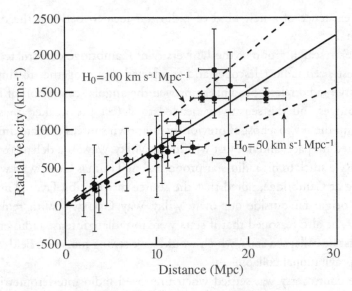

Figure 6.8 Recent Hubble diagram The *Hubble Space Telescope* has been used to obtain these measurements of the radial velocities and distances of Cepheid variable stars in galaxies. The radial velocity is given in units of kilometers per second, denoted as km s^{-1}, and the distance is in units of a million parsecs, or Mpc, where 1 Mpc is equivalent to 3.26 million light-years. These data have been used to measure the expansion rate of the Universe, also known as the *Hubble constant* and designated H$_0$, obtaining H$_0$ = 75 ± 10 km s^{-1} Mpc^{-1}. [Adapted from Wendy L. Freedman *et al.*, "Final Results From the *Hubble Space Telescope* Key Project to Measure the Hubble Constant," *Astrophysical Journal* **553**, 47–72 (2001).]

beginning. Moreover, astronomers have gazed out to the remote edges of the observable Universe, where cosmic objects can move nearly as fast as light.

Radio Stars, Radio Galaxies, Quasars and Supermassive Black Holes

When astronomers first looked out at the radio sky, they found that the brightest emission coincided with the Milky Way, so it was natural to suppose that they were stars. But because the nearest and visibly brightest stars, other than the Sun, did not exhibit detectable radio emission, and because some discrete radio sources were a hundred million times brighter than the Sun at radio wavelengths, the English radio astronomer Martin Ryle argued for the existence of a new kind of radio star located between the optically visible stars.[46]

Ryle's research group at the University of Cambridge constructed arrays of modest-sized radio telescopes and connected them together to simulate a single large radio telescope. This improved the angular resolution of discrete radio sources and the sensitivity needed to detect the weaker ones. They opened up the sky to a host of previously unknown sources that are uniformly distributed in space, and not just within the Milky Way. At a debate over their origin, Ryle stuck to his radio-star proposal, but Thomas Gold, who was then lecturing at Cambridge, noted that the source distribution favored an extra-galactic origin far outside our own Milky Way Galaxy.[47] With remarkable foresight, he also reasoned that if stars were to radiate intense radio emission they must be collapsed stars, which would have strong magnetic fields magnified by gravitational collapse.

The controversy was settled when improved radio interferometric techniques were used to refine the position of the intense radio source Cygnus A, setting the stage for the crucial optical identification that would transform our knowledge of the radio Universe. Armed with the accurate radio position, two German-born American astronomers, Walter Baade and Rudolph Minkowski, used the giant 5.0-meter (200-inch) optical telescope on Mount Palomar, California to find the visible-light counterpart of Cygnus A in 1954.[48] It is an elliptical galaxy of redshift 0.057, which meant that it is receding from us at 5.7 percent of the velocity of light. And its distance, of about 750 million light-years, was inferred from the linear relationship between redshift and distance, known as Hubble's law.

Once this distance had been established, the enormous absolute radio luminosity of Cygnus A was realized. It wasn't just emitting a faint crackle and hiss, but instead a colossal, shattering roar, like a lion instead of a household cat. Every second, Cygnus A emits as much power in radio waves as a million million, or 10^{12}, Sun-like stars radiate in visible starlight. It turned out to be a new type of radio galaxy, whose radio luminosity is comparable to its optical one.

The news that Cygnus A is a galaxy, and not a star, embarrassed and humiliated Ryle. When he saw the identifying photographs, at a scientific conference, he threw himself on a nearby couch — buried his face in his hands — and wept.[49] But he soon recovered his composure, realizing that the radio galaxies could be used to probe the distant Universe and provide tests of different cosmological models. And Ryle's group did just that, abandoning the radio star hypothesis and using their comprehensive and definitive Cambridge radio surveys to demonstrate strong evolutionary effects over cosmic times.[50] And the world forgave Ryle's earlier mistake, for he was awarded the Nobel Prize in Physics in 1974, primarily for his development of the aperture synthesis of radio telescopes.

When interferometeric measurements showed that the angular extents of some other intense radio sources were much narrower than radio galaxies, it was for a time believed that the first true radio stars of small physical size had been discovered. Allan Sandage used the 200-inch Mount Palomar telescope to look in the direction of one of them, 3C 48 — the 48th source in the third Cambridge radio-source catalogue, and there was no galaxy to be found. Instead, Sandage discovered a bright blue object, no bigger in angular size than a star, and with a totally confusing line spectrum that was unlike anything ever seen before.

The key to the mystery was provided when the Moon happened to pass in front of another bright radio source, 3C 273, in 1962. The radio astronomer Cyril Hazard, then at the University of Sydney, realized that a careful timing of the disappearance and reappearance of the occulted radio source would establish a precise position, since the location of the Moon's edge is known accurately for any time.

So Hazard and his colleagues used the occultation method to show that 3C 273 is a double radio source, one component of which apparently coincides with a blue stellar object.[51] This coincidence prompted Maarten Schmidt to confirm the identification and obtain an optical spectrum using the 200-inch Mount Palomar telescope, which indicated a completely unexpected and exceptionally high recession velocity of 0.16 percent of the velocity of light.[52] When he told his colleague Jesse Greenstein about the discovery, Greenstein produced a list of emission line wavelengths for 3C 48, and within minutes they had found that it is rushing away with an even faster motion at 37 percent of the velocity of light.

When these velocities are used to infer distances using Hubble's law, it is found that 3C 48 and 3C 273 are located at distances of billions of light-years. And when their observed brightness is combined with these distances, the intrinsic visible-light power is comparable to that of 10 million million, or 10^{13}, Sun-like stars.

Since the bright objects appeared star-like in visible light, they became known as quasi-stellar radio sources, a term that was soon shortened to quasars.

The quasars had, in fact, been ignored as stars on optical photographs for years. But a casual inspection of the optical sky would never have led to the discovery of quasars. About 3 million stars look brighter than the brightest quasar, 3C 273.

What accounts for the extraordinary power of the remote radio galaxies and quasars? They are likely energized by a compact, supermassive black hole. Its powerful gravity pulls in surrounding stars and gas, forming a flat, orbiting accretion disk that spirals into the black hole. As proposed by Martin Rees, who was later elevated to England's House of Lords as Baron Rees of Ludlow, powerful magnetic fields generated by the rotating black hole turn the whirling accretion disk into an enormous dynamo.[53] It uses the hole's rotational energy to accelerate charged particles and squirt them out in diametrically opposite directions along the rotation axis at about the speed of light. They continuously feed the two radio lobes commonly found symmetrically placed from the center of radio galaxies and quasars.

The classic example is M 87, a giant elliptical galaxy whose central spinning disk of hot gas indicates that a super-massive black hole resides at its center. A one-sided jet of gas emerges nearly perpendicular to the disk, and stretches out into one of the two lobes of the radio galaxy Virgo A, numbered 3C 274 in the Cambridge survey (Fig. 6.9). The motions of bright knots in the jet indicate that they are traveling outwards at about half the speed of light. And Very Long Baseline Interferometry observations with widely separated radio telescopes reveal that M 87's jet emerges from a region at most six light-years across, most likely harboring the super-massive black hole that produces the jet.

To power the youthful activity of a quasar by accretion, there has to be about one solar mass per year of gas flowing into the black hole. So billions of stars or the equivalent amount of gas must be consumed as a quasar or radio galaxy evolves over the course of billions of years. The supply dwindles away over time and the activity dies down, but the black hole does not disappear. Most galaxies probably contain supermassive black holes at their center. The ones in the older, nearby galaxies are the starving remains of former quasars, with a dwindling supply of material that once fed a higher rate of activity. They are found in ordinary nearby galaxies, like Andromeda and the Milky Way, whose cores are old surviving fossils of former quasars.

Astronomers decipher the history of our expanding Universe by watching massive stars explode into oblivion.

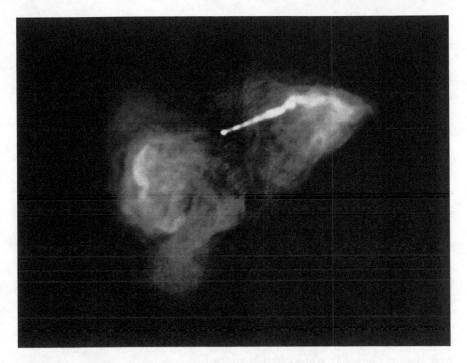

Figure 6.9   Radio jet The bright radio source Virgo A, also designated 3C 274, coincides with M 87, a giant elliptical galaxy located in the Virgo cluster of galaxies at a distance of about 50 million light-years. This radio map, made with the Very Large Array, shows two elongated lobes, one on either side of the center of M 87. The most intense radio emission comes from a jet that emerges from the core of the galaxy and stretches some 8,000 light-years into one of the two lobes. The observed high-speed motion of bright knots in the jet implies that its radio-emitting electrons are traveling at nearly the speed of light. Observations of the rates at which stars and gas clouds revolve within the central core of M 87 indicate that it contains a compact massive object, most probably a supermassive black hole of about 3 billion solar masses. (Courtesy of NRAO/AUI/NSF.)

Exploding Stars, the Supernovae

At the end of its bright, shining life, an entire star can explode and spew out its insides like phosphorescent javelins, seeding space with ingredients for the next generation of stars. As an example, the Sun, Earth and the rest of our Solar System formed 4.6 billion years ago from interstellar material that had been enriched by previous generations of massive stars that were born, lived and perished in explosions within the Milky Way. Such exploding stars also occur in other galaxies, where for a few weeks they can outshine all the rest of the galaxy. They have been given the name *supernovae*, because of their exceptional brightness.

In 1934 Walter Baade, an astronomer at the Mount Wilson Observatory, and the Swiss astronomer Fritz Zwicky, who had moved to the nearby California Institute of Technology in 1925, communicated to the United States National Academy of Sciences a remarkable pair of papers on super-novae.[54] In one paper, they demonstrated that the enormous total energy emitted in the supernova process corresponds to the complete annihilation of an appreciable fraction of the star's mass. The other publication discussed their prediction that a supernova explosion accelerates charged particles to high energies, most likely accounting for the energetic, cosmic-ray particles that rain down on the Earth from all directions. They were right on both counts.

Fritz was quite a character. Eccentric, aggressive and independent Professor Zwicky had the dangerous habit of telling everyone just what he thought of him or her. He was always out "to show those bastards," often meaning his colleagues at the California Institute of Technology, or Caltech for short, and once called them "spherical bastards," because, he said, "they were bastards any way you looked at them."

Such abrasive and outspoken behavior didn't endear him to his fellow human beings, but Zwicky didn't care — he knew he was smarter than most of the Philistines he had to work with. And his quick intelligence and irascible personality made him a formidable and indefatigable interrogator of German scientists after World War II (1939–1945), including Wernher von Braun and others associated with the German V-2 rockets.

You might say that Fritz had some kind of personality disorder, but that's what universities help protect — the exceptionally bright who don't always get along with "normal" people. And in Zwicky's case it paid off. He reasoned that supernovae ought to be exploding all the time in remote galaxies, given their immense number and the many billions of stars they each contain. So Fritz began a patrol of the night sky with a 3.5-inch camera that he mounted on the roof of a building at Caltech, accompanied by the laughter of his faculty colleagues. Much to his dismay, he did not find any supernovae for the first two years, but in 1937 he detected one.[55]

It took decades before astronomers found out what makes a star explode, and it turned out that there are two ways to detonate the explosions. Both types are symptoms of advancing stellar age, and brilliant one-way trips to

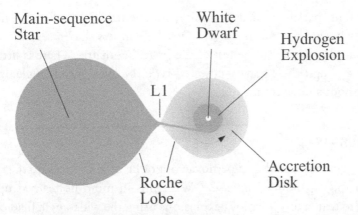

Figure 6.10 **Type I supernova** The strong gravitational attraction of a white dwarf can pull a nearby main sequence star into an elongated shape. When sufficient amounts of hydrogen in the outer atmosphere of the companion star spills over onto the white dwarf, the entire star can explode into a type I supernova.

complete destruction, but they originate in different kinds of stars and are detonated in separate ways. One type can occur in a close binary system, with a white dwarf, the shrunken dense remnant of a former low-mass star, circling another nearby ordinary star (Fig. 6.10).[56] When the nearby companion star expands, as the result of its normal evolution, hydrogen from its outer atmosphere spills onto the white dwarf, compressing and heating the star and adding mass to it. As soon as the white dwarf reaches its maximum possible mass, it can't take any more and is pushed over the edge of stability into explosion. It has become a type I supernova that shines with the light of billions of Suns, and no star is left behind.

Since every one of these nuclear-powered type I explosions come from identical stars at their maximum mass, with explosions triggered under very similar conditions, they are expected to display a uniformity in their maximum light output and the shape of their light curve. This makes them very useful as a "standard candle" for measuring the distances of very remote galaxies, and to thereby determine the pace of cosmic expansion.

In the other kind of supernova explosion, the type II variety, a very old and massive star blows up all by itself. It follows the creation of an iron core within an evolving, massive supergiant star. Since nuclear reactions cannot continue in the iron core, there is nothing left to heat the star and to generate the motions that are needed to support it. The iron core collapses under its own weight into a neutron star or black hole in less than a second, and an explosion blows away all the rest of the in-falling matter. There is little uniformity amongst these gravity-powered type II explosions of single, massive stars without a close companion.

Breaking Free

Walter Baade realized that supernovae might be used to measure the distant parts of the expanding Universe.[57] With this in mind he teamed up with Zwicky to search for supernova explosions using the wide-angle field of view and high speed of a Schmidt telescope, which could survey large areas of the sky in a reasonable time. Its inventor, Bernhard Schmidt, had worked at the Hamburg Observatory with Baade, and in 1935 Zwicky traveled to Hamburg to visit Schmidt. Upon his return, Zwicky and Robert Millikan convinced George Ellery Hale to allocate funds for an 18-inch Schmidt, in preparation for the 200-inch reflector that the Rockefeller Foundation had endowed.

The sky patrol with the new Schmidt telescope was a collaborative effort between Zwicky, Baade, and another German astronomer, Rudolph Minkowski, who had left a Professorship at the University of Hamburg in 1935 to begin a twenty-five year career at the Mount Wilson Observatory. Zwicky and Josef Johnson identified supernovae with the Schmidt telescope; Baade used the Mount Wilson reflectors to study the shape of their rising and declining light intensity, known as the *light curve*; and Minkowski used these telescopes to obtain their line spectra.[58]

By 1941 about 50 supernovae had been detected in the supernova sky patrol, and that year, Minkowski announced that there are basically two kinds of supernovae, designated as type I and type II, based on whether or not lines of the element hydrogen appear in their spectra.[59]

The sky patrol was interrupted by the Second World War (1939–1945), which also destroyed the supernova collaboration. Zwicky incorrectly accused Baade, who was an "alien" German citizen, of being a Nazi, and threatened to kill him if he showed up on the Caltech campus. From then on, Baade refused to be left alone in a room with Zwicky.

Not much happened with the supernova search for nearly half a century, when improvements in technology enabled astronomers to use some of the explosions to probe the depths of our Universe. The digital age had arrived and computers could be used with charge-coupled detectors, abbreviated CCDs. The electronic detectors collect almost 100 percent of the incident light, as compared with about 1 percent for photographic emulsions, which can make a small telescope with a CCD as powerful as a large telescope using photographic plates.

It was a time of big astronomy and big science, and the new surveys of the supernovae were big enterprises. It lasted for decades, and involved groups and teams of astronomers from all over the world using telescopes distributed around the globe in Australia, Chile, the United States and even the *Hubble Space Telescope*. Most importantly, the supernova search had a purpose beyond just watching stars explode. The inquisitive astronomers wanted to use them to find out more about what kind of expanding Universe we live in.

By collecting and recording supernova light from both nearby and distant galaxies, out to large redshifts and half-way across the observable Universe, the astronomers could measure expansion speeds out to a large fraction of the speed of light and see how the rate of cosmic expansion has changed since the light was emitted from distant supernovae. Just about everyone expected that the Universe would slow down as it continued to expand, due to the attractive gravitational force of matter within it. This meant that supernovae at great distances, which emitted their light long ago, should appear to be receding faster than those nearby.

Astronomers therefore attempted to compare the expansion velocities of very distant supernovae with those in nearby galaxies, expecting to determine the mass density of the Universe. Well, they didn't see the expected, and to everyone's surprise, observations of supernovae in remote galaxies showed that the Universe is speeding up, expanding at a quickening pace, accelerating outward and running away. The galaxies have severed the bonds of gravity, and will never return.

It took nearly 20 years of work by several teams to arrive at these conclusions, and it was a difficult road to follow.[60] In the early 1980s, it was found that a particular subclass of type I supernovae, designated type Ia, exhibit amazingly uniform line spectra and light curves indicating a similar origin and common peak luminosity. These type Ia's are identified by both the absence of hydrogen spectral features and the presence of a silicon absorption line — type Ib does not display the silicon feature but does show pronounced helium ones.

Then doubts arose over whether or not type Ia supernovae are all the same, for some of them were 10 times more luminous at peak intensity than others, so standardized light curves had to be developed.[61,62] Two international teams were organized to observe the distant, high-redshift supernovae reliably and quickly. The first one, dubbed the *Supernova Cosmology Project*, began in 1988 at the Lawrence Berkeley National Laboratory. Under the direction of Saul Perlmutter, they created software that would allow computers to find supernovae automatically from digital images taken with a small, dedicated telescope. The equipment would automatically subtract images of the night sky taken about a month apart, in times of new Moon, and anything that remained after the subtraction was a new source of light, most likely a supernova.

By the mid 1990s Perlmutter and his colleagues from Europe, Chile and America were finding a lot of supernovae, further examining them in spectral detail using major ground-based telescopes, and, occasionally, the *Hubble Space Telescope*.

Another group, dubbed the *High-z Supernova Search*, joined the quest in 1994, employing similar techniques to the *Supernova Cosmology Project*. The "*High-z*" denotes large redshifts, z, and therefore supernovae in extremely distant galaxies. This group, led by Brian P. Schmidt of the Mount Stromlo Siding Springs Observatory in Australia and the Australian National University, included Adam G. Riess, then a graduate student in the astronomy department at Harvard University.

The motivation of the *Supernova Cosmology Project* was to measure the amount of invisible dark matter in the Universe by detecting how much it slows the cosmic expansion. At first the group reported, in 1997, that they had found just what they were looking for, but stressed that the results were not precisely established, and then came to the opposite conclusion with less uncertainty a year later.[63]

In the meantime, the *High-z* team was convinced that Perlmutter's original conclusion was wrong. The breakthrough came in 1998 when Adam Riess, of the *High-z* team and by then a post-doctoral student at the

University of California at Berkeley, declared that analysis of 16 distant supernovae indicated that the cosmic expansion has unexpectedly sped up during the past 5 billion years.[64] Saul Perlmutter's group nearly simultaneously arrived at a similar result using 42 high-redshift supernovae, with publication in the following year.[65]

The two programs had independently reached the same conclusion using different supernovae and different analytical techniques. By clocking how much the cosmic expansion has changed since the light was emitted from those distant, high-redshift exploding stars, when compared with nearby low-redshift ones, they found that the expansion is speeding up and accelerating as time goes on, rather than slowing down.

In 2011 the Nobel Prize in Physics was awarded to Saul Perlmutter, Brian P. Schmidt and Adam G. Riess "for the discovery of the accelerating expansion of the Universe through observations of distant supernovae."

The unexpected runaway expansion suggested that some unknown repulsive force permeates the Universe and counteracts the combined, mutual gravitational attraction of all the matter in the Universe, pushing the galaxies faster and faster apart. It is called *dark energy*.

Cosmic Inflation

An inflation theory may describe what happened in the first fraction of a second of the expanding Universe. The novel idea was introduced by the MIT astronomer Alan H. Guth in 1981,[66,67] and further developed at Stanford University by Andrei D. Linde within a year.[68] During inflation the Universe was driven apart by a repulsive gravity, unlike the attracting kind we are used to, and operating on a very small scale in both space and time. Owing to its inherent instability, the burst of inflation soon decayed away and came to an end, in a time far less than one second, releasing its remaining energy into heat and material particles. This accelerated expansion in the first miniscule moments of the Big Bang, this inflation, most likely obliterated all evidence of previous events, in a day without a yesterday.

The cosmic inflation theory was challenged by physicists at Princeton and Harvard Universities for its current lack of definitive, observed predictions,[69] which might be compared to stirring up a hornet's nest. Particle physicists and cosmologists at MIT and the University of California, Berkeley responded to the challenge with specific quantitative predictions that match observations of the cosmic microwave background radiation.[70,71]

Regardless of the observational verification, the inflation idea has some imaginative consequences. As proposed by Alex Vilenkin at Tufts University,[72] and independently by Linde,[73] cosmic inflation indicates that multiple invisible Universes may have emerged from nothing, and other potentially unseen Universes could be waiting to arrive. If this is the case, our observable Universe might be but a small component within a vast assemblage of other Universes that we cannot see. They are together known as the *Multiverse*.

Each Universe could start with its own Big Bang, with different settings to Nature's fundamental constants and natural laws. Although any individual Universe, including our own, may live and die, the Multiverse is supposed to be forever. The eternal, self-reproducing Universes could just keep on arising from the vacuous nothing, like bubbles in the foam of a river.

All of these imaginary Universes are highly speculative. They may always remain unobservable and unverifiable, eternally unknown and unknowable, so their possible existence might never be tested, even if the cosmic inflation theory is confirmed in our own Universe.

Part II
Nothing Stays the Same

7. Natural History of the Stars

"I thought of a maze of mazes,
of a sinuous, ever growing labyrinth
which would take in both past and future
and would somehow involve the stars."

Jorge Louis Borges (1962)[1]

New Stars

For at least two thousand years, astronomers, hunters, mariners, and anyone else familiar with the brightest stars must have been amazed by a nova, or "new star," that would appear suddenly at a point in the sky where no star had been previously seen. For a few days or weeks the nova might be one of the brightest stars in the dark night sky. But then it would begin to fade away, and in about a month it would disappear back into invisibility, often without a trace.

Sometimes the new star would become so bright that it was easily visible even in full daylight. The emperor's astronomers in the Sung dynasty of China recorded one of them on July 4, 1054, near the constellation now known as Taurus, the Bull. The Chinese chronicles indicate that the new star became as bright as Venus, could be seen during the daytime for three weeks, and remained visible in the night sky for 22 months. These temporary residents were called guest stars or visiting stars. They would appear suddenly and then depart abruptly, like uninvited guests.

More than four centuries passed before the Oriental records again noted the unheralded appearance of brilliant guest stars, and this time they shook the very foundations of European thought. As Aristotle taught, the stellar Heavens were supposed to be eternal, pure, changeless, incorruptible and perfect, quite unlike anything on Earth. Yet, in a span of just 32 years two new stars appeared, remained fixed in the Heavens for about a year each, and then disappeared from view. Tycho Brahe witnessed the first visitor in 1572; it appeared in the constellation Cassiopeia with a light that was brighter than the planet Venus. Then Johannes Kepler spied another one as it lit up the Heavens in 1604.[2]

The bright new stars of 1572 and 1604 could be observed by anyone in the Earth's Northern Hemisphere. The English poet John Donne most likely referred to the 1604 nova with:

> "Who vagrant transitory Comets sees,
> Wonders, because they are rare; but a new star,
> Whose motion with the firmament agrees,
> Is miracle, for there, no new things are."[3]

Once thought to be eternal, some stars arose in the darkness and then disappeared back into it, with an apparent beginning and ending. Where had these strange and unexpected stars come from, and why did they suddenly vanish from sight after shining so brightly for months? Had they always been there and suddenly became visible, and how far away were they located? As John Donne had pointed out, comets could suddenly appear out of nowhere, remain visible for a few months, and disappear. But comets moved across the celestial background, and the novae remained fixed in the distant realm of the stars.

The incorruptibility of the Heavens was at first defended by assigning the novae to a region between the Earth and the Moon, where change was permissible. But Tycho Brahe's meticulous observations of the nova of 1572 and Galileo Galilei's and Johannes Kepler's observations of the nova of 1604 placed them far beyond the Moon and within the supposedly eternal and changeless celestial realm.[4]

The die had been cast, and there was no getting around it. At least one star residing within the celestial Heavens did come and go. Although uncommon, such a change set the stage for the ensuing realization that all stars have a history. Like humans, the stars are born, live and die. But since only a few of the thousands of the stars visible with the unaided eye had been observed to change, their lifetimes just had to be exceedingly long, much greater than those of human beings.

William Herschel's telescopic discoveries in the 18th century indicated that stars congregate into clusters, and that there are nebulous regions in the night sky that contain no stars. This indicated to him that stars and star clusters formed in the past, are changing, and will eventually perish over vast spaces and enormous times.

William Herschel, Musician Turned Astronomer

William Herschel, one of England's most famous astronomers, started life in Germany, never received formal training in astronomy or physics, and earned his living as a musician for the first half of his life. He was born in Hanover on November 15, 1738, and named Friedrich Wilhelm Herschel. At the age of 14 he joined his father Isaac and older brother Jacob as an oboist in the regimental band of the Hanoverian Foot-Guards and accompanied the regiment to England in 1755, as a precaution against the threat of a French invasion. The following year they returned home, where the French defeated the Foot-Guards. Wilhelm and Jacob fled to England in the autumn of 1757.

Wilhelm arrived in London at the age of 19, nearly penniless, but soon found work as a free-lance musician. After nearly ten years of this wandering life, William, as he had become known, settled down as the organist of the Octagon Chapel in Bath, a fashionable health resort for the aristocracy, and brought his sister Caroline to Bath, to keep house, launch a brief musical career, and help with his astronomical observations.[5]

For the next decade, Herschel became a musician by day and an astronomer by night. He was totally self-educated in astronomy, by reading books and learning to make excellent telescopes. Leisure, comfort, meals, and sleep were sacrificed to grinding and polishing metal mirrors of ever-increasing size that would collect more light. With unrivaled skill, William was able to build telescopes with the largest known mirrors, which could resolve greater detail on bright cosmic objects or bring fainter ones into view. He could then see what no one else had ever seen before, either with unaided eyes or with any other person's telescope during his lifetime.

At a time when most astronomers were preoccupied with the planets, which moved against a backdrop of fixed stars, Herschel wanted to find out more about the stars themselves. In the fall of 1779, at 41 years of age, he therefore embarked on a systematic review of the stars using a telescope of his own making.

Herschel was looking for two stars; one bright and the other dim, but so close together that they appeared to be single even with the best telescope other than his own. He found hundreds of these "double" stars; even the North Star turned out to be one. At the time, no one knew how far away the stars were, and Herschel hoped that the two stars could be used to infer the distance of the brightest one by an annual parallax change in the apparent positions of the two stars. The technique was flawed because it assumed the fainter member of the pair would be much farther away than the brighter one. When Herschel returned to some of these newfound double stars later in his career, he discovered that some pairs had changed angular separation from each other, which meant that they were close companions joined together in orbital motion around a shared center and not widely separated in distance from the Earth.

After about a year and a half of this work, William discovered, on March 13, 1781, a totally unexpected, non-stellar object, which he initially thought was a comet.[6] But he was using very high telescopic power, and saw right away that this object had an observable disk, unlike any star. "The goodness of my telescope," he recalled, "was such that I perceived its visible planetary disk as soon as I looked at it."[7] Several months of observation by Herschel and others showed that the moving object is a major planet that orbits the Sun beyond Saturn, at about twice Saturn's distance.

Herschel proposed that the new planet be named *Georgium sidus*, Latin for "*George's star*," in honor of King George III, England's reigning monarch.[8] After considerable dispute, the new planet eventually became known as *Uranus*, after Urania a Greek goddess and muse of astronomy, which is the name used today.

It was the first planet to be discovered since the dawn of history, and it made Herschel famous. By the end of 1781 he had been elected a Fellow of the Royal Society, was awarded one of its gold medals, and would soon be supported by the King.

Just eight years after Herschel's discovery of the unknown planet, a newly discovered, heavy element was designated uranium in honor of the finding of the new world. Then in a little more than a quarter century, the young English poet John Keats likened his own discovery of Homer's poetry to Herschel's sighting of a new planet and to an explorer finding new lands or seas:

> "Then felt I like some watcher of the skies
> When a new planet swims into his ken;
> Or like stout Cortez when with wondering eyes

> He stared at the Pacific — and all his men
> Looked at each other with a wild surmise —
> Silent upon a peak in Darien."[9]

At almost exactly midway in his 83-year lifetime, William became Astronomer to the King with a royal pension, which permitted him to give up music as a career and devote full time to astronomy, except when called upon to demonstrate his telescopes to the Royal Family and guests.

For the rest of his life, Herschel continued with what he did best — constructing telescopes with larger mirrors and greater light gathering power than ever obtained before, and using them to survey the Heavens.

By late in 1783, Herschel had built a telescope that would be his main observing instrument. It had a mirror of 18.7-inches (0.47-meters) diameter and a 20-foot (6.1-meter) focal length. He used it to count the number of stars in various directions and thereby establish the "Construction of the Heavens."

With extraordinary perseverance and zeal, Herschel would sit at his telescope from sunset to sunrise, always in the open air, often in freezing temperature, and in great physical discomfort. Night after night and year after year he continued to sweep the sky in an organized and methodical way, often with the help of his faithful sister Caroline.

It took them two decades, from 1783 to 1802, to complete a survey of the entire sky visible from their home, which resulted in the discovery of over 2,500 nebulae and star clusters, most of them never seen before.[10]

William Herschel's reports of vast numbers of previously unknown nebulae and star clusters suggested that they change as the result of relentless gravity, a universal agent of transformation.

The Stars Gather Together

Most scientists of William Herschel's time agreed with the ancient Greek concept of immortal, unchanging stellar Heavens. To avoid the inevitable gravitational disruption of the stars, Herschel adopted Isaac Newton's proposal that the known Universe began with a near-uniform distribution of stars, which "the great Author" would preserve, but his observations also suggested that the stars are not regularly spread through the space they reside in. After over two decades of surveying the "Construction of the Heavens,"

Herschel discovered more than a thousand star clusters, which indicated to him that gravity had begun to draw some stars together.

The French astronomer Charles Messier had already prepared a list of 103 bright star clusters and nebulae in order to avoid confusion in his hunts for comets.[11] Published in final form in 1781, the year that Uranus was discovered, the Messier list includes some of the most widely studied objects in the Universe, all designated by the letter "M" followed by the number in the Messier catalogue. Herschel naturally looked at the bright Messier objects using his superior telescope. It had twenty five times the light collecting area and five times the resolving power of the one Messier used, which meant that Herschel could detect much fainter and finer details.

All but one of Messier's twenty-nine star clusters were listed by Messier as "round nebulae," which might be confused with a comet without a tail, but they were not yet resolved into stars. Herschel was able to detect stars in some of them, and coined the term *globular star cluster* for these objects. Prominent examples include M 3, M 5 and M 80, which have all been well studied by subsequent generations of astronomers.

Herschel discovered that the stars are compressed together near the centers of the brightest globular star clusters, and that they are more sparsely congregated further out. Individual stars could only be picked out near a cluster's outer boundary, which gradually thinned out to the emptiness of surrounding space (Figs. 7.1, 7.2). So how, he wondered, did the globular star clusters get this way?

To Herschel, their spatial distribution portrayed a *Natural History of the Heavens* in which star systems have a beginning, development, and end. Any dense irregularity in the original distribution of stars would, as the result of its greater gravitational pull than normal, attract adjacent stars into a common center. As time went on, the central stars would be drawn toward each other by their mutual gravitational attraction and move ever closer together, so they evolved from widely dispersed conditions to the tightly packed globular star clusters. This would account for their round shapes, as well as their dense centers and more rarefied peripheries. It would also explain the surrounding empty space from which the stars had apparently been swept up and pulled out of.

Moreover, Herschel proposed, star clusters might clock the growth of the Universe. Each cluster would be as old as the time it took for it to achieve its present shape. Dispersed star clusters would be relatively young in comparison to older, compact ones, and the star clusters could collectively serve as

Figure 7.1 Globular star cluster Hundreds of thousands of stars are held together by their mutual gravitational attraction in M 80, a globular star cluster whose age is about 14 billion years. It is about as old as the observable Universe and significantly older than our Sun whose age is about 4.6 billion years. (A *Hubble Space Telescope* image courtesy of NASA/AURA/ STScI/Hubble Heritage Team.)

"laboratories of the Universe." They resembled cosmic fossils with their life history imprinted into their shape, and signaled the eventual decay of the known stellar Universe, the Milky Way.

An Enchanted, Luxuriant Garden in the Sky

According to Herschel, the nebulae and star clusters display the history of a heavenly garden, a life cycle of stellar systems in which gravity acts as a universal agent of change. He therefore wrote in 1789:

> "[The Heavens] now are seen to resemble a luxuriant garden, which contains the greatest variety of productions, in different flourishing beds; and one advantage we may at least reap from it is that we can, as it were, extend the range of our experience [of them] to an immense duration."[12]

Figure 7.2   Stellar beauties A *Hubble Space Telescope* image of stars in the globular star cluster NGC 6397, which has an estimate age of about 12 billion years. (Courtesy of NASA/ESA/Harvey Richer, U. British Columbia.)

In this interpretation, star systems were like plants at various stages of growth and decay, and you could "judge the relative age, maturity or climax from the disposition of their component parts." The stable, clockwork Universe of Newton was thus being replaced by a Cosmos that is always changing and has a history. The exact time scale remained unknown, but cosmic change would have to happen over enormous times, owing to the vast spatial scale of the stellar Universe.

Erasmus Darwin, an English poet, physician, and naturalist, celebrated Herschel's vision of stellar change in a popular poem entitled *The Botanic Garden*, published in 1791. It described an enchanted garden that included the wonders of stars and planets that exploded into being once God, the Creator, started it all, including:

"So late described by Herschel's piercing sight,
Hang the bright squadrons of the twinkling night.

— Roll on, ye stars! Exult in youthful prime,
Mark with bright curves the steps of time."[13]

There would come a time, Darwin proclaimed, when the light of stars
would be extinguished into the dark, for:

"Flowers of the sky! Ye to age must yield,
Frail as your silken sisters of the field!
Star after star from Heaven's high arch shall rush,
… to one dark center fall,
And death and night and chaos mingle all!"[14]

Then in November 1790 Herschel came across "a most singular
phenomenon," a bright star enveloped by a faint luminous nebula (Fig. 7.3).

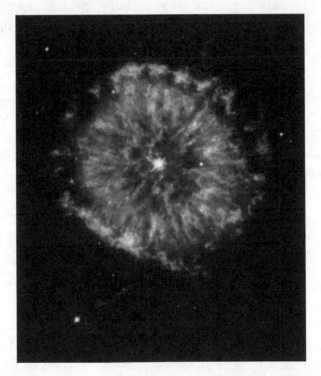

Figure 7.3   Planetary nebula When a Sun-like star uses up its nuclear fuel, the star's center
collapses into an Earth-sized white dwarf star while its outer gas layers are ejected into space.
(A *Hubble Space Telescope* image of NGC 6751 courtesy NASA/STScI/AURA/Hubble
Heritage Team.)

The "nebulosity about the star," he noted, "is not of a starry nature," and its self-luminous material was "more fit to produce a star by its condensation than to depend on a star for its existence."[15]

Near the end of his career, Herschel arranged his observations of nebulae and star clusters into a proposed sequence of advancing age. It was supposed to demonstrate how gravity dominates stellar life and sculpts cosmic objects into different configurations and shapes or forms. The process begins with faint, milky nebulosity covering considerable areas of the sky. Ongoing gravitational attraction draws these diffuse clouds into smaller, concentrated patches, which grow more regular in shape and brighten in the center. As gravity continues to work its magic, these in turn become smaller and more compact stellar systems, and the aging process ends with the magnificent globular star clusters.

It was, Herschel thought, like observing babies, children, and adults to see how human life unfolds. Furthermore, when most of the stars have ripened into globular form, the Milky Way must break up and cease to be a stratum of stars. After all, it was already starting to fragment under the destructive power of gravity.

And if the Milky Way was falling apart, it must have had a beginning, or in Herschel's words: "Since the breaking up of the parts of the Milky Way affords a proof that it cannot last forever, it equally bears witness that its past duration cannot be admitted to be infinite."[16] He associated this starting point with its Divine Creation.

Creation of the Stellar Universe

Like many educated people in his day, Herschel believed the first stars were fashioned by a Divine Creator, whose power and glory can be seen in the heavenly, star-clad night sky. To him, it provided evidence for, and even proves: "the existence of a first cause, the infinite author of all dependent things."[17]

William nevertheless spoke so little of his faith that after his death his son John wanted it to be "distinctly understood that my father … was a sincere believer in, and worshipper of, a benevolent, intelligent and superintending Deity, whose glory he conceived himself to be legitimately forwarding by investigating the magnificent structure of the Universe."[18]

This structure was a testimony to the Creator of the Heavens, as in Joseph Addison's *Hymn*, written a quarter century before Herschel's birth:

> "The Spacious Firmament on high,
> With all the blue, Ethereal Sky,
> And spangled heavens, a shining frame,
> Their great Original proclaim."[19]

Herschel's contemporary, the composer Joseph Haydn, similarly celebrated the stars in his *The Creation*. And about a quarter century after that, Friedrich Schiller's beautiful *Ode to Joy* was set to music in Ludwig van Beethoven's *Ninth Symphony* to proclaim our Creator, who above the stars must dwell.[20]

8. How the Sun and Planets Came into Being

"He, who through vast immensity can pierce,
See worlds on worlds compose one Universe,
Observe how system into system runs,
What other planets circle other suns,
What vary'd being peoples ev'ry star,
May tell why Heav'n has made us as we are."

Alexander Pope (1732)[1]

The Nebular Hypothesis

The presently accepted nebular hypothesis for the origin of the Solar System supposes that the Sun and planets were created together during the gravitational collapse of a rotating, interstellar cloud. The spinning gaseous solar nebula kept collapsing until its central regions became so concentrated and hot that the Sun began to shine. The planets formed at the same time within a flattened, rotating proto-planetary disk centered on the contracting proto-Sun (Fig. 8.1).

The earliest known mention of the concept was by the Swedish scientist, theologian, and Christian mystic Emanuel Swedenborg in his *Principia* in 1734. The German philosopher Immanuel Kant extended Swedenborg's ideas in 1755, when he reasoned that a rotating gaseous nebula would collapse and flatten due to gravity. Kant described how the stars might have become systematically arranged in the Milky Way through the gravitational collapse of a large rotating nebula, and how the Solar System could have originated by the pulling together of a much smaller rotating nebula.[2]

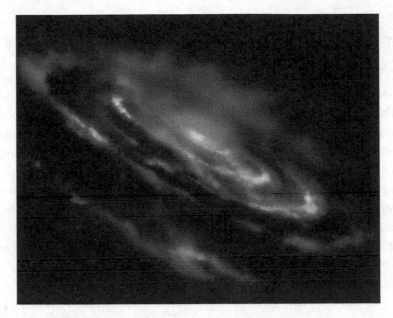

Figure 8.1 Formation of the Solar System According to the nebular hypothesis, the Sun and planets were formed at the same time during the collapse of a rotating interstellar cloud of gas and dust that is called the solar nebula. The center collapsed to ignite the nuclear fires of the young Sun, while the surrounding material was whirled into a spinning disk where the planets coalesced. (Courtesy of Helmut K. Wimmer, Hayden Planetarium, American Museum of Natural History.)

Kant was a devout Christian who believed in God, the Divine Creator. The regular, ordered development of the Solar System did not happen by chance, he argued, and it was God who imposed that order, that design, on the material world in His act of Creation. To Kant, it was proof of God's existence.

So Kant was not casting doubt on Who created the Heavens, but was, instead, humbly considering how God might have gone about it. He also proclaimed that all the fixed stars are suns with similar planetary systems; all formed and produced "out of the smallest particles of the elementary matter that filled empty space — that infinite receptacle of the Divine Presence."[3] He additionally proposed that the stars are not eternal, that they had a beginning, and that they must all perish. But we need not lament their passing, Kant supposed, any more than those of flowers or insects, "all demonstrations of the Divine Omnipotence," and "while nature thus adorns eternity with changing scenes, God continues engaged in incessant creation in forming the matter for construction of still greater worlds."[4]

The French astronomer and mathematician Pierre-Simon Laplace independently proposed the nebular hypothesis for the formation of the Solar System in the same year as Kant did; and additionally proposed that the rotating nebula shed a succession of rings from which the planets formed. The prolific Laplace was elected to the French Académie des Sciences at the age of 24. The Academy had never, observed its secretary, "received from so young a candidate in such a short time so many important papers on varied and difficult topics as the sequence submitted by Laplace.[5] His achievements in these and later years included gravitational astrophysics, the Laplace transform, spherical harmonics, potential theory, the proof of the method of least squares, probability theory, and the speed of sound.

Kant and Laplace led quite different lives. Kant never married, and in his entire 79-year life never traveled more than a few kilometers outside of his hometown, in Königsberg, the capital of Prussia at that time, and now Kaliningrad, Russia. In contrast, at the age of 39, Laplace married Marie-Charlotte, just eighteen and a half years old, at the Saint-Sulpice in Paris. Within a few years, they had two children, most of Laplace's contributions to astronomy had come to a close, and he became involved in French political affairs.

Both astronomers realized that the nebular hypothesis would account for the regular arrangement of the orbits of the major planets and their satellites. Any origin theory, Kant noted, must explain the fact that the planets all orbit the Sun within a common plane, which coincides with the equatorial plane of the Sun, and that they all revolve in approximately circular orbits in one direction that is the same as that of the Sun's rotation. Kant attributed this "conformity in the direction and position of the planetary orbits" to the one "material cause through which they were all set in motion."[6] That cause, he proposed, is the contraction of a diffuse, rotating cloud of gas particles under the influence of gravity. The greatest increase in mass would occur at the center, which would become the Sun, and the surrounding material would consist of independent particles in circular motion within a plane about the central body.[7]

Laplace Draws Attention to the Nebular Hypothesis

Although Kant's speculations were prescient, it is not clear how influential they were. At about the same time that his *Universal Natural History* was printed, the publisher became bankrupt and his holdings were impounded. Only a few copies of Kant's book reached the public, and his work does not seem to have been noticed by his contemporaries, including William Herschel. In the following century others nevertheless drew attention to Kant's astronomical speculations, which incidentally contained no mathematical details.

It was Pierre-Simon Laplace who in 1796 first drew public attention to a much more popular, but similar, process for the origin of the Solar System, while apparently unaware of Kant's related work. Laplace's account appeared at the very end of his *The System of the World* (*Exposition du Système du Monde*), which was published in five editions over almost thirty years, with successive modifications to its famous ending pages on the formation of the Solar System.[8]

These pages begin with a review of the regular orbital arrangement of the planets and their satellites, which to Laplace could not be accidental. The major planets all revolve around the Sun in nearly the same plane, and the main satellites move about their planet in the same direction and plane that it orbits the Sun. Moreover, the planets and satellites all turn about their axes in the same direction and plane as their orbital motion, which coincide with the Sun's direction of rotation and equatorial plane.

Although astronomers had previously noted these startling alignments, Laplace was the first to show they could not arise by chance. If the known planets and satellites were haphazardly thrown together into randomly oriented configurations, the probability that they would have the same directions and planes of orbital and rotational motion is exceedingly small and unlikely. Laplace showed that even if millions upon millions of Solar Systems were made in this random way, only one would be expected to look like our own.

Faced with this dilemma, Laplace looked for a solution other than chance alignment. It is provided by the nebular hypothesis in which the planets formed out of the collapsing, rotating extended atmosphere of the young Sun. Laplace supposed that the planets were formed within "zones of vapors," successively thrown off from the newborn Sun, and that Saturn's rings illustrate the earliest stages of the formation of the much larger system of planets.

We now turn to Laplace's friendship with Napoléon Bonaparte and their oft-quoted discussion about God, including Herschel's on-the-spot account of the exchange.

Laplace, Herschel, Napoléon Bonaparte and God

Immediately after he seized power as First Consul in the coup d'état of 1799, Napoléon Bonaparte named Pierre-Simon Laplace the Minister of the Interior, but after only six weeks in the government, Bonaparte's brother, Lucien, replaced Laplace. Much later, while exiled to Saint Helena, Napoléon wrote his reminiscence of the short-lived appointment, noting that Laplace "sought everywhere for subtleties, conceived only problems, and in short carried the spirit of the infinitesimal into administration."[9]

Realizing that he should retain the allegiance of the eminent scientist, Napoléon appointed Pierre-Simon to the French Senate, named him to the Legion of Honor, and ennobled him as Count of the First French Empire. His wife joined the court of Napoléon's sister, princess Elisa, where they focused on the world of fashion.

In return, Laplace dedicated the third volume of his *Mécanique Céleste* to Napoléon and included an adulatory dedication to him in the *Théorie Analytique des Probabilités*.

Then in 1814, when it became evident Napoléon's empire was falling and that the monarchy would be restored, Laplace offered his services to the Bourbons. His change in allegiance was affected by his realization that Napoléon had overextended himself, putting the French nation at risk, as well as the fact that Laplace's son Émile was endangered while participating in the fighting at the Eastern Front.

Laplace was also offended by Napoléon's reaction to the death of Laplace's only daughter Sophie-Suzanne in childbirth. On returning from the rout in Leipzig, Napoléon told Laplace that he had grown thin, to which he replied: "Sire, I have lost my daughter." Napoléon replied "Oh! That's not a reason for losing weight. You are a mathematician; put this event in an equation, and you will find that it adds up to zero."[10]

In another celebrated exchange, Napoléon looked over Laplace's *Exposition du Système du Monde*, and commented to Laplace: "Newton has spoken of God in his book. I have already gone over yours and I have not found this name a single time." To this, Laplace responded: "Citizen First Consul, I had no need of that hypothesis (Je n'avais pas besoin de cette hypothèse-là)."[11] But Laplace was not claiming that God does not exist. He was instead stating that the gravitational interaction of the planets always remains stable, and that God did not need to intercede to break the laws of Nature and retain their equilibrium, as Newton had once proposed.

In 1802, William Herschel traveled to Paris and had several meetings with Laplace. During that visit, Napoléon Bonaparte had an audience with Herschel, Laplace, and others. The First Consul asked Herschel about his findings on the Construction of the Heavens, and addressed Laplace on the same subject. When asked to explain how that structure came about, Laplace attempted to show it was the result of natural laws. Napoléon argued against that view, and in Herschel's account of the meeting:

> "The difference was occasioned by an exclamation of the First Consul, who asked in a tone of exclamation or admiration (when we were speaking of the extent of the sidereal heavens): 'And who is the author of all this!' Mons. De Laplace wished to shew that a chain of natural causes would account for the construction and preservation of the wonderful system. This the First Consul rather opposed. Much may be said on the subject; by joining the arguments of both we shall be led to 'Nature and nature's God'."[12]

In other words, it is God's Nature that exhibits the causal, ordered structure described by natural laws that were set into place by the Creator.

There were many more sublime discoveries about planetary worlds in ensuing times, and many of them were related to the nebular hypothesis of Kant and Laplace.

The Quest for New Worlds

If the nebular hypothesis applies to the formation of all the stars, then they should all be surrounded by spinning protoplanetary disks or orbiting planets. The first evidence for these planet-forming disks was obtained in the early

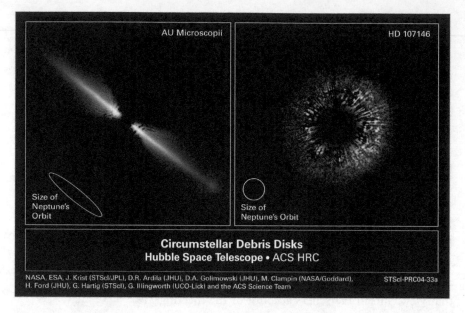

Figure 8.2   Dusty disks Starlight is reflected from thick disks of dust that might still be in the process of forming planets. They surround the stars Au Microscopii (*left*) and HD 107146 (*right*) that are respectively thought to be 12 million years old and between 50 million and 250 million years old. [*Hubble Space Telescope* images courtesy of NASA/ESA/STScI/JPL/ John Krist – STScI/JPL (*left*) and David Ardila – JHU (*right*)]

1980s with instruments aboard the *InfraRed Astronomical Satellite*, and decades later the *Spitzer Space Telescope* used its powerful infrared vision to detect hundreds of stars that may be surrounded by planet-forming disks. In fact, the youngest nearby stars are usually found embedded in the dense clouds of interstellar gas and dust that spawned them.

The *Hubble Space Telescope* has been used to discover flattened disks of dust swirling around many young stars (Fig. 8.2). They suggest that the nebular hypothesis applies to them, and material in the disks is expected to coalesce into full-blown planets if it hasn't already done so. The Atacama array of radio telescopes that operates at millimeter wavelengths is now being used to resolve the protoplanetary disks around nearby stars (Fig. 8.3).

Individual planets are almost always too small and too faint to be seen directly by the reflected light from their parent star. Their presence has only recently been inferred from their gravitational effects on the motions of the

Figure 8.3 Protoplanetary disk The Atacama Large Millimeter/submillimeter Array (ALMA) has been used to obtain this image of planet formation around a young, Sun-like star HL Tau, which is probably no more than a million years old. Emerging planets may have swept their orbits clear of dust and gas. [Courtesy of ALMA (NRAO/ESO/NAOJ); C. Brogan, B. Saxton (NRAO/AUI/NSF).]

star they revolve around, or when they chance to pass in front of a star, momentarily blocking the star's light when viewed from Earth. Such extrasolar planets, which orbit around stars other than the Sun, are called *exoplanets*.

To detect their presence, astronomers had to look for the subtle compressing and stretching of starlight as an unseen planet tugged on a star and pulled it first toward the Earth and then away. This causes a periodic shift of the stellar radiation to shorter and then longer wavelengths (Fig. 8.4). To detect the effect, astronomers have to observe the wavelength of a well-known spectral feature, called a line, and measure the periodic shift of its wavelength.

But an orbiting planet produces an exceedingly small variation in the wavelength of spectral lines emitted from its star. So the effect could not be detected until the starlight was dispersed into fine wavelength intervals and collected by electronic charge-coupled detectors. And since no single line shift is significant enough to be seen, computer software had to be written to add up all the star's spectral lines, which shift together, combining them over

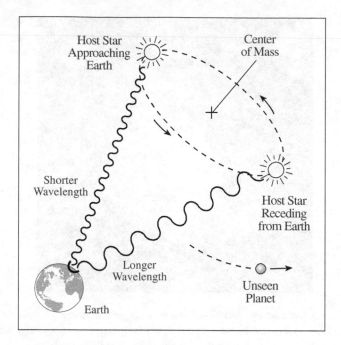

Figure 8.4 **Starlight shift reveals invisible planet** An unseen planet exerts a gravitational force on its visible host star, which periodically approaches and recedes from Earth. This motion changes the wavelength of the starlight seen from Earth through the Doppler effect, and reveals the presence of the planet orbiting the star.

and over again at all possible regularities, or orbital periods, with continued comparison to non-moving laboratory spectral lines.

It took decades for astronomers to develop these complex and precise instruments. Then, in the 1990s, the time was ripe, and two Swiss astronomers from the Geneva Observatory in Switzerland, Michel Mayor and Didier Queloz, discovered the first planet that orbits an ordinary star, the faintly visible, Sun-like star 51 Pegasi, only 48 light-years away from Earth in the constellation Pegasus, the Winged Horse. As Mayor described it, the occasion "happened like a dream, a spiritual moment."[13]

They had detected the back-and-forth Doppler shift of the star's light with a regular 4.23-day period, measured by a periodic change of the star's radial velocity of up to 50 meters per second (Fig. 8.5).[14] To produce such a quick and relatively pronounced wobble, the newfound planet had to be large, with a mass comparable to that of Jupiter, which is 318 times heftier

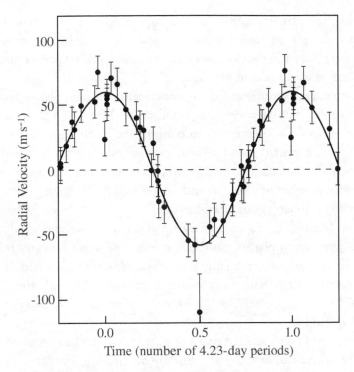

Figure 8.5 **Pioneering planet discovery** An unseen planet of about half the mass of Jupiter produces a periodic 4.23-day variation in the observed radial velocity of the star 51 Pegasi measured in units of meters per second, or m s^{-1}. The invisible companion is in close orbit around the star, at a distance of just 0.05 of the mean distance between the Earth and the Sun. [Adapted from Michael Mayor and Didier Queloz, "A Jupiter-mass Companion to a Solar-type Star," *Nature* **378**, 355–359 (1995).]

than Earth, and moving in a tight, close orbit around 51 Pegasi at a distance of only 0.05 AU, where 1 AU is the mean distance between the Earth and the Sun. By way of comparison, Jupiter is about 100 times further away from the Sun than the newfound planet is from the star 51 Pegasi. No one had anticipated that a giant planet would be orbiting so close to its star.

The intense radiation and powerful winds of a newly-formed star were expected to keep any nearby material from gathering together into a large planet, explaining why Jupiter and the other giant planets were formed far from the Sun in the cold, outer precincts of our Solar System. But it was a good thing for planet hunters, for the large mass of a giant world would produce a more pronounced velocity change than the smaller mass of an Earth-sized planet and the close orbit meant a short orbital period that might be detected in weeks instead of years.

Less than two weeks after the announcement of a giant planet circling 51 Pegasi, two American astronomers, Geoffrey W. Marcy and R. Paul Butler, used their own past observations to confirm the result, and now that they knew that giant planets could revolve unexpectedly near a star, with short orbital periods, they used powerful computers to re-examine their previous observations of other nearby stars and announced the discovery of two more Jupiter-sized companions of Sun-like stars.[15]

From then on, anyone could look up at the night sky and say that there are definitely unseen planets out there, orbiting perfectly ordinary stars that are now shining brightly in the night sky. *Time Magazine* celebrated the event, on Februrary 5, 1996, with a cover story entitled: "Is Anybody Out There?" They might have also used: "Are We Alone?" So far, no one knows the answer to either question.

After astronomers realized that a large planet could be so near to its star, they knew where and how to look. And by monitoring thousands of nearby Sun-like stars for years, they have now found thousands of planets revolving about other nearby stars, many of them massive planets that have been dubbed "hot Jupiters" because of their size and proximity to the intense stellar heat. They are much too hot and too close to their star for life to survive or water to exist.

Finding Habitable Exoplanets

From a human perspective, the most interesting exoplanets will be those as small as the Earth, in circular orbits at just the right distance from the heat of a Sun-like star to permit liquid water to exist on the planet. Scientists call this location a *habitable zone*. At closer distances, the water would all be boiled away, and at more remote distances it would be frozen solid. A planet in this zone could be inhabited, but that does not mean that life does reside on it.

Such a planet produces too small a gravitational perturbation of their star to be detected by the velocity method with existing technology, so the transit method has to be used to detect them. That is, the planet's orbit has to be in just the right orientation for it to periodically cross directly in front of, or transit, its host star.

The orbital size can be calculated from the period of the repeated transit and the mass of the star. From the orbital size and the brightness of the star, the planet's temperature can be calculated. This information would tell us if the planet is warm enough for liquid water to exist on its surface.

The *Kepler* mission was specifically designed to detect planets comparable to the Earth in size or smaller, and located at or near this habitable zone. By measuring the brightness of 100,000 stars, it has detected the periodic dimming of starlight produced when thousands of planet candidates pass in front of their star, and that is quite an accomplishment. A transit by an Earth-sized planet in the habitable zone will produce a small change of only about 0.0001, or 1/10,000, in the star's brightness lasting for 2 to 16 hours.

Most of these planet candidates are orbiting nearby stars that are smaller and cooler than our Sun. Since these stars are less luminous than the Sun, the habitable zone is closer to the star than the Earth is from the Sun, and planets within it have orbital periods that are less than our year, which means that they can be recognized in relatively short observation times of a few years. As an example, seven Earth-sized planets have been found orbiting the ultra cool Trappist-1 star, and three of these planets are located in the star's habitable zone. It is located 39.5 light years from the Sun in the constellation Aquarius.

The *Kepler* planet candidates require follow-up verification observations with the world's best telescopes on the ground and in space. Atmospheres have even been detected for some of these exoplanets, but no one knows if any of them are inhabited by living things. That possibility remains as speculative as it has always been.

It has long been thought that there might be living things on planets other than the Earth, either in our Solar System or orbiting stars other than the Sun. The arguments have been extensively documented in two lengthy books — by Steven J. Dick from the ancient Greeks to Kant, and by Michael J. Crowe from Kant to the beginning of the 20th century.[16] The scope of our book does not include such all-embracing considerations, so lets just say the idea that there is life out there beyond the Earth just will not go away, and take up the thread of our discussion where we last left astronomers, in the mid-18th century.

According to the nebular hypothesis, the formation of planets should be a natural result of star formation, and planets should encircle most stars. Moreover, some of these planets could be populated with diverse living beings. Immanual Kant and Pierre-Simon Laplace both speculated about life on other planets in our Solar System.[17,18] And William Herschel was no less reluctant than Kant or Laplace in populating planetary worlds with imaginary beings. A year before his discovery of Uranus, Herschel speculated that the observed features on the Moon indicate that it is inhabited. Later, he proposed that living creatures also reside on the Sun. The sunspots, he noted, could uncover and reveal a dark and solid Sun where other beings might live, beneath its fiery atmosphere.

Inhabitants of the Sun was a bit too much. As the great English astronomer A. S. Eddington remarked: "It cannot be denied that he [William Herschel] was given to jumping to conclusions in a way which, when it comes off, we describe as profound insight, and when it does not come off, we call wildcat speculation."[19]

Eventually astronomers determined the physical conditions on the Moon, Sun and planets in our Solar System, including their temperatures and atmospheres, and concluded that life, as we know it, is not likely to be found on them. They instead focused on then unanswered questions about how the Sun and other stars shine.

9. The Ways Stars Shine

"What is the past, after all, but a vast sheet of darkness
in which a few moments, picked apparently at random, shine."

John Updike (1963)[1]

The Substance of the Stars

Before astronomers could understand how the Sun and other stars produce their heat and light, they needed to know what they are composed of. In a brilliant doctoral dissertation, published in 1925, the American astronomer Cecilia H. Payne used observations of spectral lines to show that hydrogen is by far the most abundant element in the outer atmosphere of the Sun and many other stars.[2] But she could not believe that the composition of stars differed so enormously from that of the Earth, where hydrogen is found rarely, so Cecilia mistrusted her understanding of the hydrogen atom. Prominent astronomers of the time also did not think that hydrogen could be the main ingredient of the stars, and this may have played a role in her considerations.

The Danish astronomer Bengt Strömgren, an expert on stellar interiors, next calculated the hydrogen content inside stars, assuming they are chemically homogeneous, and showed in 1932 that their observed luminosities require that the entire star, and not just its outer atmosphere, must be predominantly composed of hydrogen.[3]

We now know that hydrogen is the most abundant element in the Universe, and that there was nothing wrong with Miss Payne's observations. The Earth just does not have sufficient gravity to retain hydrogen gas in its atmosphere for any length of time. It either evaporated away while the Earth was forming, or became locked into water or surface rocks at that time.

Helium, the second-most abundant element on the Sun, is so rare on the Earth that it was first discovered in the Sun, when the French astronomer Pierre Jules César (P. J. C.) Janssen found an unidentified yellow emission line in the solar spectrum observed during the solar eclipse of 18 August 1868.[4]

The English astronomer Joseph Norman Lockyer observed the same yellow line at about the same time and was subsequently knighted for this discovery. It was probably not until the following year that Lockyer convinced himself that the yellow line could not be identified with any known terrestrial element, and named the solar element "helium" after the Greek Sun god, *Helios*, who daily traveled across the sky in a chariot of fire drawn by four swift horses.[5]

Helium was not found on Earth until 27 years after its discovery in the Sun, when the Scottish chemist Sir William Ramsay discovered its spectral signature in a gaseous emission given off by a heated uranium mineral cleveite.[6] Today, helium is used on Earth in a variety of ways, including inflating party balloons, and in its liquid state to keep sensitive electronic equipment cold. Though plentiful in the Sun, helium is almost non-existent on the Earth. It is so terrestrially rare that we are in danger of running out of helium during this century. Japanese scientists have proposed that helium may be extracted and returned from the Moon's surface, where the terrestrially rare element has been implanted by winds from the Sun.

Altogether, 92.1 percent of the atoms of the Sun are hydrogen atoms, 7.8 percent are helium atoms, and all the other heavier atomic elements make up only 0.1 percent. In contrast, the main ingredients of the rocky Earth are the heavier elements like silicon and iron, which explains the Earth's high mass density — about four times that of the Sun, which is about as dense as water.

It was the English astronomer and devout Quaker A. S. Eddington who noticed that the conversion of hydrogen into helium might provide stellar energy from the mass lost in the process.

A. S. Eddington, Seeking the Truth

Arthur Stanley Eddington, known as A. S. for short, was born on December 28, 1882 in Kendal, England, the second child and only son of Quaker parents, themselves from Quaker families. His father, headmaster of the Friend's School in Kendal, died of typhoid when Arthur was only 2 years old, and his mother was left to bring up her children with little income.

The family moved to Weston-super-Mare where Arthur was educated at home before spending three years at preparatory school. In 1893 he entered the Bryn Melyn School, where he excelled in mathematics and English literature. His talents enabled him to earn a scholarship to Owens College in Manchester before he had reached the age of 16. After four years at the College, where he won all the honors it had to offer, A. S. was awarded a scholarship to Trinity College at the University of Cambridge. In 1904 he became the first ever second-year student to achieve top honors in its mathematical exams, known as the *Tripos*.

Two years later, at the age of 23, Eddington was appointed Chief Assistant to the Astronomer Royal at the Royal Observatory in Greenwich, England, where he investigated stellar movements and the structure of the stellar system.

Eddington was elected as Plumian Professor of Astronomy at the University of Cambridge in 1913, to replace George Darwin who had just died. The following year, as the result of another person's demise, Eddington also became director of the Cambridge Observatory, and therefore responsible for both observational and theoretical astronomy at the University. He held these positions with great distinction for the next thirty years.

Eddington's tenure at Cambridge did not have a happy beginning, for it coincided with the early stages of World War I (1914-1918). As a Quaker pacifist, he claimed conscientious objector status and refused to serve in the war. When called up for conscription, he publicly declared, before a Cambridge Tribunal, that his religious beliefs could not support the call to slaughter other human beings. The tribunal refused to accept his deferment, and only the timely intervention of the Astronomer Royal and other high profile figures kept Eddington out of prison and on his way to observe the total solar eclipse of 1919 described in Chapter 2.

Eddington never married, avoided romantic entanglements, and lived in Cambridge, England with his mother and sister, and later with his sister alone who survived him. They provided a home suitable to his temperament and work, and thereby helped A. S. dedicate his life to understanding the stars.

With exceptional physical insight and mathematical ability, he explained how stars move, gather together, support themselves, pulsate, curve nearby space-time, depend on mass, light up nearby space, and stay hot inside and shine.[7] His elegant book entitled *The Internal Constitution of the Stars* summarizes his pioneering considerations of the structure, composition, energy source, and evolution of stars.[8]

A. S. also had his lighter side. He enjoyed solitary cycling through the English countryside in the spring and fall, and was addicted to solving the crossword puzzles in *The Times* and *The New Statesman and Nation*. He was an avid reader of mystery novels, and once likened the process of understanding stars to analyzing the clues in a crime. Eddington wrote wonderful accounts of astronomy with humor, literary allusions, analogies, metaphors, and similes that the educated public loved and still does.[9]

There was a continual interaction and overlap of Eddington's scientific career and religious outlook.[10] He identified himself as both a Quaker and an astronomer, and believed that seeking the truth through mystical experience is fundamental to both religion and science. There is no absolute, certain knowledge, he thought, in either realm, and he urged everyone to participate in a continued quest for truth in all ways.

In *Science and the Unseen World*, Eddington wrote: "In science as in religion the truth shines ahead as a beacon showing us the path; we do not ask to attain it; it is better far that we be permitted to seek it…. As truly as a mystic, the scientist is following a [pure and holy] light. … And so in the light walking and abiding, these things may be fulfilled in the Spirit, not in the letter; for the letter killeth, but the Spirit giveth life."[11]

In describing why he believed in God, he declared: "The desire for truth so prominent in the quest of science is a reaching out of the spirit from its isolation to something beyond, a response to beauty in Nature and art, and [the] Inner Light of conviction and guidance."[12]

And when a cry goes up from the human heart about the mystery of our existence, asking what is it all about or why do humans exist? Eddington replied that life is "about a spirit in which truth has its shrine, with potentialities of self-fulfillment in its response to beauty and right."[13] "I cannot believe that human beings exist simply in order to breed more millions of human beings," he declared. "The human race must be aiming, in some way, at something finer."[14]

Eddington dealt with the unseen worlds within stars, and he turned to another invisible world to seek truths beyond science. "For the rest," he wrote, "the human spirit must turn to the unseen world to which it itself belongs." It is a different, transcendental perception, and you cannot apply natural laws to it "anymore than you can extract the square root of a sonnet."[15]

A. S. never supposed that he, or any other astronomer, was infallible. He viewed astronomy as an ongoing, open-ended pursuit of the truth. It is an unfinished quest, always changing, endlessly approximating, forever improving, continually moving forward, and never final. Taking part in this magnificent exploration is what mattered to Eddington, not the ultimate correctness of any given hypothesis.

When narrating the story of Daedalus and Icarus in 1920, he therefore wrote:

> "In ancient days two aviators procured to themselves wings. Daedalus flew safely through the middle air across the sea, and was duly honored on his landing. Young Icarus soared upwards towards the Sun till the wax melted which bound his wings, and his flight ended in fiasco. In weighing their achievements perhaps there is something to be said for Icarus.... So, too, in Science, cautious Daedalus will apply his theories where he feels most confident they will safely go; but his excess of caution cannot bring their hidden weaknesses to light. Icarus will strain his theories to the breaking point till the weak joints gape."[16]

So, lets fly on with Eddington and Icarus (Fig. 9.1) to peer within the Sun and other stars.

Figure 9.1 Icarus The mythological Icarus seems to be pushing against the downward pull of gravity, trying to break free, soar into a bright blue sky, and set the human spirit free. Icarus' red heart symbolizes love, for: "He who loves also soars, runs and rejoices; he is free and nothing can restrain him." [A paper-cut out by the Henri Matisse, courtesy of Succession Matisse, Paris.]

How does the Sun Shine?

As Eddington acknowledged, his investigations of the hidden interior of the Sun began with the American astronomer Jonathan Homer Lane's realization in 1870 that gravitational compression within the Sun and other stars make them exceptionally hot inside.[17] Lane was the first to investigate the Sun as a gaseous body, and he assumed that the gas pressure of hot, moving particles supports the mass of the Sun. The central temperature can be inferred by assuming equilibrium between the gas pressure and the inward gravitational pull at the star's center, which occurs at a central temperature of 15.6 million degrees kelvin. A star that is more massive than the Sun produces greater compression at its center, and a higher central temperature is required to support it.

Eddington showed that all of the energy of any star has to be released deep down inside its high-temperature core, and that no energy is created in the cooler regions outside the core. A star is energized in its central regions and that energy is transported out to a star's visible disk.

The power that must be produced within the Sun is just amazing. When we measure the total amount of sunlight that illuminates and warms the Earth, and extrapolate back to the Sun, we find that it is emitting an enormous power of 385.4 million, million, million, million, or 3.854×10^{26}, watts. In just one second, the Sun expends more energy than humans have used since the beginning of civilization.

So what produces that prodigious energy? In the mid-nineteenth century, the German physicist Hermann von Helmholtz proposed that it is due to the Sun's gravitational contraction. As gravity slowly pulls the solar material inward, compressing and heating it up, the hot solar gases would keep the Sun shining. As subsequently shown by the Irish physicist William Thomson, later known as Lord Kelvin, slow gravitational collapse might supply the Sun's energy for about 100 million years.[18]

The astonishing thing, which was not realized at the time Thomson wrote his influential article, is the Sun's durability. It has lasted much longer than he envisioned. Fossil evidence indicates that sunlight has been sustaining life on Earth for about 4 billion years, so the Sun has been kept hot inside and radiated away energy at its enormous rate for billions of years. Only nuclear reactions can fuel the Sun's fire and make it shine so brightly for so long.

Moreover, as A. S. first pointed out, the situation is dramatically worse for the giant stars. In 1917, he wrote: "If contraction is the only source of energy, the giant stage of a star's existence can scarcely exceed 100,000 years."[19] He instead proposed that hydrogen is transformed into helium inside stars, with the resultant mass difference released as energy to power them.

At the end of World War I in 1918, the chemist Francis W. Aston returned to the University of Cambridge, where he invented the mass spectrograph and used it to show that the mass of the helium nucleus is slightly less massive, by a mere 0.7 percent, than the sum of the masses of its ingredients. When a new helium nucleus is bound together, a tiny amount of mass is lost and released in the form of energy. Eddington rightly concluded that this could supply the Sun's current luminous output for an estimated 15 billion years.

In his far-sighted 1920 essay on the internal constitution of the stars, Eddington wrote:

"Certain physical investigations in the past year make it probable to my mind that some portion of sub-atomic energy is actually being set free in the stars. F. W. Aston's experiments seem to leave no room for doubt that all the elements are constituted out of hydrogen atoms [protons] bound together with negative electrons. The nucleus of the helium atom, for example, consists of four hydrogen atoms [protons] bound with two electrons. But Aston has further shown conclusively that the mass of the helium atom is less than the sum of the masses of the four hydrogen atoms that enter into it…. Now, mass cannot be annihilated, and the deficit can only represent the mass of the electrical energy set free in the transmutation … The total heat liberated will more than suffice for our demands, and we need look no further for the source of a star's energy."[20]

[It was subsequently found that the nucleus of the helium atom consists of two protons and two neutrons, but their combined mass is not noticeably different from that of four protons — see Chapter 3.]

Albert Einstein provided the basic idea when he showed that mass and energy are interchangeable. Every mass has an equivalent energy, just as every form of energy has an equivalent mass.[21] Energy radiated may, for example, be supplied by mass lost. [Their equivalence is expressed by the single elegant equation $E = mc^2$, where E denotes energy, m stands for mass, and c^2 denotes the square of the speed of light c. Because the speed of light is a very large number, only a small amount of mass is needed to produce a large amount of nuclear energy.]

In Eddington's view, the stars are the places where the elements are assembled by fusion. The mass of the assembled element is less than the mass of its ingredients, and the lost mass provides the energy that keeps stars hot inside and fuels their radiation. In this way, he linked some of the largest bodies in the Universe to the smallest, and joined the stars to the atoms and their nuclear constituents.

In the same article, Eddington also included the prescient statement that:

> "If, indeed, the sub-atomic energy in the stars is being freely used to main-tain their great furnaces, it seems to bring a little nearer to fulfillment our dream of controlling this latent power for the well being of the human race — or for its suicide."[22]

So it is nuclear fusion reactions in the compact, dense, high-temperature core of a star that energizes the particles there, sustaining their heat and mak-ing them move rapidly. Once the nuclear reactions begin, the sub-atomic energy that is liberated keeps the nuclei sufficiently hot to insure the continu-ation of the reactions. They are termed "nuclear" reactions because it is the interaction of atomic nuclei that powers the stars. For the Sun and the vast majority of other stars, it is protons, the nuclei of hydrogen atoms, which fuse together to make the nuclei of helium atoms.

The chains of nuclear reactions involved in transforming hydrogen into helium inside stars depend upon the mass of the star. For most stars, including the Sun, there is a direct conversion of four hydrogen nuclei, the protons, into one helium nucleus. This chain of nuclear reactions was discovered by Hans Bethe in 1938,[23] immediately following a conference in Washington, D.C. organized by George Gamow to discuss the problem of stellar energy genera-tion. For exceptionally massive stars, which are more massive than the Sun, carbon acts as a catalyst in the nuclear transformation, following the chain that was delineated by Carl Friedrich von Weizsäcker the previous year.[24]

[Carl's brother Richard von Weizsäcker (1920-2015) studied at Balliol College, Oxford, returned to serve in the German Army in World War II (1939–1945), and became President of the Federal Republic of Germany from 1984 to 1994.]

Tunneling Through the Barrier

Physicists were nevertheless convinced that protons could not overcome their mutual electrical repulsion and fuse together inside the Sun. Protons are positively charged, which means they repel one another, like identical ends of two magnets. The force of repulsion between like charges becomes larger and larger as they are brought closer and closer to one another. Even in a high-speed, head-on collision at the enormous central temperature of the Sun, two protons are not moving fast enough with sufficient kinetic energy to overcome their mutual charged repulsion and come together. But A. S. was certain that nuclear energy fueled the stars, and remarked that: "The helium which we handle must have been put together at some time and some place. We do not argue with the critic who urges that the stars are not hot enough for this process; we tell him to go find *a hotter place.*"[25]

George Gamow had provided the resolution of this paradox in 1928 when he demonstrated how α particles tunnel out of the nucleus of a radioactive atom (Chapter 3). Of course, a star is not radioactive, and it is a place where elements are being synthesized instead of spontaneously decaying. But a similar tunneling process, or barrier penetration, occurs the other way around at the center of stars like the Sun. It means that a proton has a very small but finite chance of occasionally moving close enough to another proton to tunnel through or under its electrical barrier.

Only exceptionally fast protons, moving much faster than average, can merge together when they collide, and that rarely happens. Even with the help of tunneling, the average proton has to make about ten trillion trillion, or 10^{25}, collisions before nuclear fusion can happen. That explains why the Sun does not expend all its nuclear energy at once, and shines for billions of years. If all the protons at the center of the Sun were moving fast enough to fuse together, the star would explode like a colossal hydrogen bomb.

Properties of the Stars

Eddington's contributions to our understanding of the stars depended upon the observations of dedicated astronomers who were measuring fundamental stellar properties such as distance, mass, and luminosity.

The measurement of the distance of a nearby star, other than the Sun, involves careful scrutiny of two stars that appear close together in the sky and are not members of a binary system (Fig. 9.2). The distance of the nearer star can be determined by comparing its position to that of the more distant one when viewed from opposite sides of the Earth's orbit, or from a separation of twice the mean distance between the Earth and the Sun, which is known as the *astronomical unit* and abbreviated AU. The angular change in position is known as the *annual parallax*,

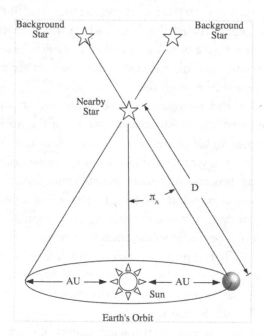

Figure 9.2   Annual parallax When a distant and nearby star are observed at six-month intervals, on opposite sides of the Earth's orbit around the Sun, astronomers measure the angular displacement between the two stars, or $2\pi_A$, which can be used to determine the distance, D, to the star.

from *annual* for the Earth's year-long-orbit and the Greek *parallaxis,* for the "value of an angle."

During the course of a year, the nearby star will seem to sway to and fro, in a sort of cosmic minuet that mirrors the Earth's orbital motion. The nearer the star is to us, the larger the annual-parallax sways. Once this parallax angle is combined with the known value of the AU, the star's distance can be established by triangulation, the geometry of a triangle.

To see the parallax effect, hold a finger up in front of your nose, and look at your finger with one eye open and the other closed, and then with the open eye closed and the closed one open. Any background object near to one side of your finger seems to move to the other side, making a parallax shift. When this is repeated with your finger held farther away, the angular shift is smaller. In other words, the more distant an object is, the smaller the observed parallax.

A nearby star will therefore display a larger annual parallax than a more distant one, and provided that all stars move at about the same velocity the closer ones will also exhibit the largest angular change of location in the sky over a given length of time. That is why Wilhelm Bessel choose the "flying star" 61 Cygni, of known large proper motion, to make the first measurement of a star's annual parallax and distance in 1838.[26] With amazing determination, Bessel measured the tiny angular separation between 61 Cygni and two close stars with the utmost care every clear night for fifteen months, usually repeating his observations at least sixteen times every night. Then in December 1838 he announced that 61 Cygni weaves back and forth by an amount that indicates the star is about 700,000 times farther away from us than the Sun is, which also meant that the dark spaces between the stars are exceptionally vast.

Another crucial parameter is the mass of a star. A direct measurement of stellar mass can be obtained from observations of the relative motion of two stars that are revolving about a common center. If the orbital period and the distance separating the two stars are measured, the sum of their masses can be determined from Kepler's third law. It turned out that there is comparatively little variation in the mass of stars, or as A. S. expressed it: "There is relatively not much more diversity in the masses of new-born stars then the masses of new-born babies."[27]

But don't belittle the mass, for a little difference can become very important. A small increase in a star's mass implies, for example, a big increase in its luminosity. Stars of lower mass have less weight pressing down on their cores, so their cores are cooler, the rates of their energy-producing nuclear reactions at these lower temperatures are slower, and the stars are dimmer. The amount of time a star shines also depends on its mass. The more massive a star is, the shorter its life. A star of greater mass is more luminous, burns its nuclear fuel at a greater rate, and uses up its available energy in a shorter time.

Eddington realized that the gas pressure of the massive, high-luminosity giant stars could not support them against the inward force of their immense gravity. If the inner temperatures of giant stars were high enough to generate a gas pressure sufficient to balance gravitation, then their luminosity would greatly exceed that actually observed. This enigma was resolved in 1917 when Eddington demonstrated that these stars are supported by radiation pressure, which increases with the fourth power of the temperature.[28] In contrast, the gas pressure is just proportional to the temperature, so if you increase the central temperature enough the radiation pressure will become much larger than the gas pressure.

At a great enough mass, the star becomes so hot inside that it is blown apart. The internal radiation pressure of such an exceptionally hot star will blow away its outer atmosphere. This explains why there are no known stars with a mass greater than about 120 times the mass of the Sun.

While investigating the unseen interiors of the luminous giant stars in 1924, Eddington also came across a mass-luminosity relation, which unexpectedly applied to all stars for which mass and luminosity had been determined (Fig. 9.3).[29] As he demonstrated, this correlation of mass and luminosity meant that every star could be treated as a gas sphere.

In deriving his mass-luminosity relation, Eddington abandoned the mathematical certainty of theoretical physics, and instead focused on equations that describe observations and have physical meaning. His major rival, James Jeans, routinely challenged such results at meetings of the Royal Astronomical Society, where he attacked Eddington for his sloppy lack of mathematical precision. In 1925, for instance, Jeans stated: "All Eddington's theoretical investigations have been based on assumptions which are outside the laws of physics."[30] When stars are instead investigated with mathematical

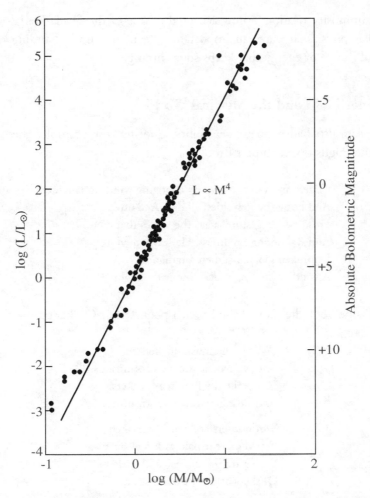

Figure 9.3 Stellar mass-luminosity relation The observed mass-luminosity relation for main-sequence stars of absolute luminosity, L, in units of the solar luminosity, $L_\odot$, and mass, M, in units of the Sun's mass, $M_\odot$. The straight line corresponds to a luminosity that is proportional to the fourth power of the mass.

rigor, he asserted, the mass-luminosity relation disappears and vanishes from the realm of physical law.

Eddington replied that mathematicians often lack the physical insight needed to understand the stars. His practical approach, rooted in observations as well as mathematics, has usually withstood the test of time, and the mass-luminosity relation is now accepted as one of the fundamental results of stellar physics. It plays an important role in our understanding of the internal source of energy that keeps stars shining.

An Inner Light and the Mystical World

Eddington also had a poetic sensibility, quoting, for example, these lines from the English poet Rupert Brooke:

> "There are waters blown by changing winds to laughter
> And lit by the rich skies, all day. And after,
> Frost, with a gesture, stays the waves that dance
> And wandering loveliness. He leaves a white
> Unbroken glory, a gathered radiance,
> A width, a shining peace, under the night."[31]

In the same book, he quoted the English poet Arthur O'Shaughnessy:

> "We are the music-makers,
> And we are the dreamers of dreams,
> Wandering by lone sea-breakers,
> And sitting by desolate streams;
>
> World losers and world forsakers,
> On whom the pale moon gleams:
> Yet we are the movers and shakers
> Of the world for ever, it seems."[32]

Eddington believed in the inner light of human beings, and in a mystical world outside space and time, while also becoming one of the most accomplished stellar astronomers who ever lived. When he died in November 1944, at the age of 62, the great American astronomer Henry Norris Russell wrote: "The death of Sir Arthur Eddington deprives astrophysics of its most distinguished representative."[33]

Our text now turns to Russell himself, who was also a devout astronomer committed to faith and science, God and Nature, in his life and work.

10. The Paths of Stellar Life

> "Two roads diverged in a yellow wood,
> And sorry I could not travel both ...
> I took the one less traveled by,
> And that has made all the difference."
>
> Robert Frost (1916)[1]

A Lifelong Princetonian with Cosmic Power

Henry Norris Russell was born on October 25, 1877, at Oyster Bay, Long Island, half of Puritan and half of lowland Scots stock. His father, Alexander Russell, was a pastor of the local First Presbyterian Church, who had graduated from the Princeton Theological Seminary in New Jersey. Henry's mother, Eliza Norris, met his father when he was staying at her family home in Princeton. The couple had two other sons, Gordon and Alexander, born in 1880 and 1883, respectively.

Henry spent most of his school years in Princeton, where he stayed at the Norris family home, and returned each summer to Oyster Bay. At age 12 he entered the Princeton Preparatory School and enrolled in the local College of New Jersey a month shy of 16. He graduated at age 19 — at the head of his class and with extraordinary honors; by that time the college had changed its name to Princeton University.

The bright young man immediately joined the fledging graduate program at Princeton University, which led to his Ph.D. in mathematical astronomy in 1900 with a thesis on the way Mars perturbs the orbit of the asteroid Eros. But it wasn't easy. Henry spent all his strength in completing his degree, and had to take two years off to rest. He recuperated from the breakdown in trips with his mother to the island of Capri and other parts of southern Italy.

It was a privileged and sheltered life. Money was never an issue. During his student days, Henry did not earn his own way, even in part. Housing and meals with the Norris family were free, and because he was a minister's son tuition at Princeton University was waived. His Aunt Ada Norris provided for any other fees, and legacies to his mother from her parents had made their family economic position fairly comfortable, at least with careful spending which came by tradition from the New England housewives who "feared dirt, debt, and the Devil, and nothing else."[2]

College life fitted well with Henry's evangelical beliefs. Most of the Professors were church members, chapel attendance was compulsory, and the students were happy and knew it.[3] As an adult, he did not drink coffee or tea, smoke, or indulge in any alcoholic drinks beyond a bit of sherry before dinner. It is even said that his thesis about the asteroid Eros caused him embarrassment, owing to the name *Eros*, the Greek god of love.

The Princeton Professor Charles A. Young stimulated Russell's interest in astronomy through his lectures and his *Manual of Astronomy*. Young also connected his astronomical and religious views, writing that astronomy "reveals the glory and majesty of the Creator, the eternal, omniscient and all pervading God."[4] He also quoted from the *Bible*, stating in his lectures that: "The Heavens declare the glory of God; and the firmament sheweth his handiwork."[5]

After graduation from Princeton, Russell spent nearly two years at the Cambridge University Observatory, where he photographed eclipsing binary stars and helped use them to determine their physical properties.

In 1905 Russell returned to Princeton University, after he persuaded Woodrow Wilson, then President of Princeton, that they needed a good astrophysicist just like himself to discover how stars evolve, how the Universe reached its present form, and what will become of it.[6]

And that is precisely what Russell built his career on. He became a Full Professor of Astronomy in 1911, and Director of the Princeton University Observatory a year later. He spent nearly his entire professional life at Princeton University, living in the same old Norris-family house from 1890 to 1957.

Like his mentor Professor Young, Henry Norris Russell was a devout Christian, and in later life he also wrote about the importance of non-scientific questions. Human beings, he declared, can be described by their average behavior, with an understanding of any individual that defies complete scientific determination. That is because:

> "Science answers only the question: '*How* do these things come to pass?' ...But the methods of mathematical analysis are no longer of prime importance for the question which the child asks first of all, "*Why* are things so?"[7]

Russell wrote that personal freedom and responsibility, and our capacity for good and evil, have to be taken into account when discussing our uncertain human existence, but that we should always keep in mind that "man was created to glorify God and enjoy him forever."[8]

Russell believed the Universe exists because there is a Power behind it, and as human beings "we can assuredly have relations of some sort with the Cosmic Power. .... No one feels more keenly than the student of Nature the greatness and splendor of that Power."[9] This transcendent Deity, our God, is "clear outside those limitations of space and time within which the material Universe, and we as parts of it, have our being ... a Being who is the only reason *why* there is any Universe."[10]

For this God, the distinctions of before and after do not exist. The human soul, Russell thought, outlasts death and belongs to this immortal realm. As an astronomer, he likened this soul to starlight that can persist long after a star has ceased shining. "Even if a star is dead and gone, its light lives on — undiminished, individual, immortal." The immortal human soul may also "survive indefinitely, through an unlimited time, retaining its full individuality, never becoming merged with any other personality or lost in some vague undifferentiated whole."[11]

Some of the most distant stars and galaxies may no longer exist, but their light can survive unchanged. As long as it encounters no matter, starlight can travel within empty space forever. Moreover, there is no interference between lights traveling through space from different stars. Thus, even when viewing thousands of stars, a telescope can focus on the light of any particular star and form an individual image of it.

Upon death, Russell believed, we commend our spirit to God. It's something we can stake our lives on.

Russell's student Harlow Shapley has written a fine biographical memoir about his mentor's accomplishments in stellar astronomy,[12] and David H.

DeVorkin's book-length biography is filled with rich detail about Russell's personal and scientific life, and his efforts to integrate theory into mainstream American astronomy.[13]

For 43 years, starting in 1900, Henry contributed monthly popular articles about the discoveries, status, and progress of astronomy in *The Scientific American*. This opportunity was due to Princetonian ties, through the owners of the magazine who had attended the University. Russell also wrote a widely used, two-volume textbook of *Astronomy*, coauthored with two Princeton colleagues R. S. Dugan and J. Q. Stewart. This revision of Young's *Manual of Astronomy* was first published in 1926–1927, with supplements and revisions over the next two decades. In 1935, Russell additionally wrote an influential book about *The Solar System and Its Origin*, and in 1940 he published an important monograph on *The Masses of the Stars*, with Charlotte E. Moore.

Henry Norris turned down the directorship of the Harvard College Observatory, and was content with wielding more influential power through endless correspondence and visits, from Harvard to the Lick and Mount Wilson Observatories in California, and through his *Astronomy* textbook and *Scientific American* columns.

Russell became President of the American Astronomical Society, the American Association for the Advancement of Science, and the American Philosophical Society. He was also awarded the gold medal of the Royal Astronomical Society of England, two medals of the French Academy, five other medals of American scientific societies, and numerous honorary degrees. Altogether, Henry Norris Russell was quite an eminent, influential, versatile, and accomplished astronomer, the ultimate Princetonian.

In spite of his accomplishments, in later life Russell was also tormented by deep depression, self-doubt, and indecision, and suffered from frequent periods of collapse from exhausting overwork and nervous, restless activity.

His insight to the path of stellar life depended on the dedicated work of other astronomers, many of them women.

How Stars Age

Working under the direction of Edward C. Pickering, astronomers at the Harvard College Observatory examined the photographic spectra of hundreds of thousands of stars, and determined their dominant spectral lines and temperatures. These astronomers were mainly young ladies who had studied physics or astronomy at nearby colleges for women, such as Wellesley and Radcliffe. Harvard did not educate females at that time and did not permit them on its faculty.

One of these faithful, stalwart workers was Annie Jump Cannon, who classified the spectra of roughly 400,000 stars between 1918 and 1924.[14] She distinguished the stars on the basis of the absorption lines in their spectra, and arranged most of them in a smooth and continuous spectral sequence. The hottest stars, with the bluest colors, were designated as spectral type O, followed in order of declining visible disk temperature by spectral types B, A, F, G, K and M (Table 10.1). Stars that displayed spectral lines of highly ionized elements, for example, were relatively hot because high temperatures are required to ionize atoms. Stars that displayed spectral lines of unionized hydrogen atoms would be cooler, and those with molecular lines cooler still. These spectral classifications eventually led to an understanding of the way stars evolve.

The probable course of a star's aging might be inferred by assuming it is a large, hot gaseous sphere whose heat is sustained by its contraction. August Ritter, a Professor of Mechanics at the Polytechnic University of Aachen, Germany, published a series of eighteen important papers that described such a process and proposed how it could be related to a star's spectral type. These

Table 10.1. The spectral classification of stars[a]

Class	Dominant Lines	Color	Visible Disk Temperature	Examples
O	He II	Blue	28,000–50,000	χ Per, ε Ori
B	He I	Blue-White	9,900–28,000	Rigel, Spica
A	H	White	7,400–9,900	Vega, Sirius
F	Metals; H	Yellow-White	6,000–7,400	Procyon
G	Ca II; Metals	Yellow	4,900–6,000	Sun, α Cen A
K	Ca II; Ca I	Orange	3,500–4,900	Arcturus
M	TiO; Ca I	Orange-Red	2,000–3,500	Betelgeuse

[a]An H denotes hydrogen, He is helium, Ca is calcium, and TiO is a molecule. The Roman numeral I denotes a neutral, unionized atom, the number II describes an ionized atom missing one electron, and the temperatures are in degrees kelvin.

papers did not attract much attention until George Ellery Hale, the editor of the newly formed *Astrophysical Journal* initiated the publication of an English version of the sixteenth paper in 1898.[15] Here Ritter introduced a classification of stars with rising temperature as well as a falling one. A star would become hotter while undergoing gravitational contraction, and after reaching the acme of its brilliance, the star would cool and evolve toward extinction. The peak luminous output would occur just before a star has been compressed so much that it could no longer behave as a gas. At this critical density the star could not supply more heat by further contraction, and it would just cool down and radiate the leftover heat away.

In the early 20th century it was also proposed that stars might change over time in ways suggested by their spectral lines.[16] This set the stage for Henry Norris Russell's display of stellar luminosity as a function of the disk temperature inferred from spectral lines. Russell's diagram became a primary tool for tracing the paths of stellar evolution, which is still used today.

Russell's Diagram

In 1914, Russell published his famous plot of the stellar luminosities against spectral class or disk temperature (Fig. 10.1),[17] which displays the trajectory of a star's life and demonstrates how its properties change with age. For many years this figure was called the *Russell Diagram*, but eventually astronomers realized that the Danish astronomer Ejnar Hertzsprung had previously published, in 1911, a less extensive diagram relating stellar luminosity and color for the nearby star clusters, the Pleiades and Hyades.[18] It therefore became known as the *Hertzsprung-Russell diagram*, or H-R diagram for short.

Because he had access to the parallax measurements, and therefore distances, of many stars, Russell's diagram included rare and very luminous red stars of spectral class M, as well as faint red stars of the same class M. According to Russell, the noteworthy aspect of his diagram was that most stars form a continuous slanting progression from the luminous B to faint M stars. The other notable aspect of the diagram was that the red stars were either very faint or very luminous with none observed in between at intermediate luminosity.

Russell predicted that if his survey was extended to many thousands of stars, the diagram would be represented by two lines: one descending

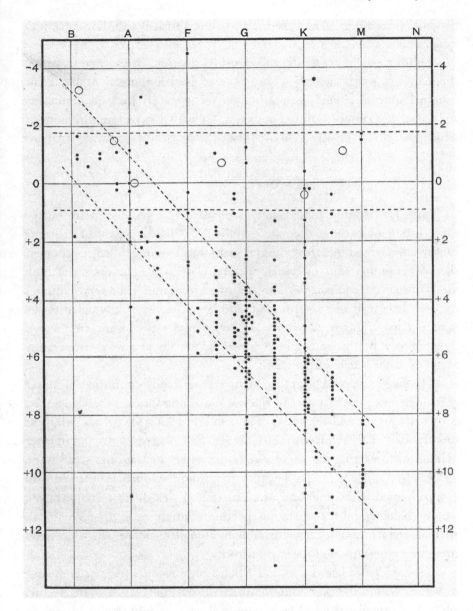

Figure 10.1   Russell's diagram In 1914 Henry Norris Russell published this diagram of the luminosity (*vertical axis*) plotted as a function of spectral class (*top horizontal axis*) for four moving star clusters. The two diagonal lines mark the boundaries of Ejnar Hertzsprung's observations of the Pleiades and Hyades open star clusters in 1911; this is now known as the main sequence along which most stars, including the Sun, are located. [Adapted from Henry Norris Russell: "Relations Between the Spectra and Other Characteristics of Stars," *Popular Astronomy* **22**, 275–294 (1914).]

diagonally from B to M and the other starting also at B and running almost horizontally.

Another way of expressing this two-fold division, Russell stated, is that there are two stellar classes — giant stars of great luminosity of about one hundred times the Sun's luminosity, and varying very little in luminosity from one class of spectrum to another, and dwarf stars of lower luminosity, which fall off very rapidly in luminosity with increasing redness.[19]

Giant and Main-Sequence Stars

As Ejnar Hertzsprung subsequently asserted: "The giant stars are indeed more luminous because they are 'swollen.'"[20] Russell's study of eclipsing binary stars showed that the luminous giant stars have much larger diameters than other fainter stars of similar spectral class, while the masses of both kinds of stars are comparable. This indicated to him that the very luminous red stars have a smaller mass density and larger size when compared to the faint red stars. This is also just what you would expect from the thermal radiation of a hot, gaseous sphere, whose luminosity at a fixed temperature increases as the square of the radius.

The large size of at least one giant star was fully confirmed by direct measurement of the diameter of the red star Alpha Orionis (Betelgeuse) by Albert Michelson and Francis G. Pease in late 1920. Michelson, who was then located at the University of Chicago, had pioneered the interference method of measuring the angular sizes of sources that are too small to be resolved by a single telescope. The radiation from the source is observed with two connected mirrors that act as an interferometer, or interference-meter, whose changing mirror-separation produces interference fringes with an effective angular resolution comparable to a single telescope with a diameter equal to the separation of the two mirrors.

George Ellery Hale, a friend of Michelson, suggested that he come out to Mount Wilson and use his interferometer to measure the size of Betelgeuse, whose large angular size had been predicted by A. S. Eddington. Michelson and Pease took the long train ride to California, and mounted their mirrors on a 20-foot (6-meter) steel beam placed at the end of the open tube of the 100-inch (2.5-meter) telescope on Mount Wilson. By measuring the mirror separation when the interference fringes disappeared, they concluded that

Betelgeuse has an angular diameter of 0.047 seconds of arc.[21] Current observations indicate a very similar angular diameter of 0.055 seconds of arc, which means that Betelgeuse is 1,180 times the size of the Sun, and has a radius comparable to the distance between the Sun and Jupiter.

A century of increasingly accurate and extensive observations have confirmed the initial characteristics of the H-R diagram (Fig. 10.2). The majority of stars, including the Sun, lie in a band that extends diagonally from the upper left to the lower right, or from the high-luminosity, high-temperature, blue stars to the low-luminosity, low temperature red stars. Russell dubbed these more numerous stars *dwarfs*, since they are smaller than the giants, but the designation is confusing. There is, for example, no observable difference between the size and luminosity of the hottest dwarf and giant stars. Astronomers now retain the designation giant stars, but use the term *main-sequence stars* for the other ones, a name suggested by A. S. Eddington in the 1920s. The stars on the main sequence are the most common type of star in the Milky Way, constituting 90 percent of its stars.

How do Stars Begin their Lives?

Before astronomers could use the Hertzsprung-Russell, or H-R, diagram to trace out a star's life, they had to determine how stars are born and where they first appear on that diagram. Henry Norris Russell had one interpretation, but he was dead wrong. In his proposal, the stars begin their lives as vastly extended, low-temperature giant stars of spectral type M. Under the influence of gravity they collapse, grow smaller and hotter, and move along the top of the H-R diagram as they age. When the star reaches the upper left-hand corner of the H-R diagram, further compressibility is no longer possible. Cooling begins and the star moves from top left to bottom along the main sequence. It decreases in luminosity and temperature, and ends up as a faint star of spectral type M.

Russell's interpretation of the H-R diagram had to be abandoned when Eddington showed in 1924 that a single mass-luminosity relation applies to all stars,[22] which led Russell to abandon his interpretation of the main-sequence stars. He instead suggested that a star's mass may be an important factor in solving the great problem of stellar evolution.[23]

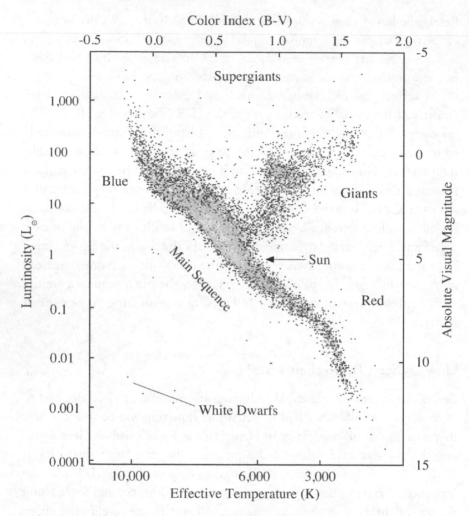

Figure 10.2 H-R diagram for nearby stars The luminosity (*left vertical axis*), in units of the Sun's absolute luminosity, $L_\odot$, plotted against the effective temperature of the star's disk in degrees kelvin, designated K (*bottom horizontal axis*) for 22,000 stars in the catalogue of the *HIPPARCOS* satellite. Most stars, including our Sun, lie along the main sequence that extends from the upper-left to the bottom-right sides of the diagram. Stars of about the Sun's mass evolve into helium-burning red giant stars, located in the upper-right side of the diagram. (Data points courtesy of ESA/*HIPPARCOS* mission.)

The Danish astronomer Bengt Strömgren subsequently concluded that it is mass and hydrogen content that determine a star's position in the H-R diagram. In 1933, he additionally proposed that studies of stars in clusters would help determine the evolutionary interpretation of the diagram.[24]

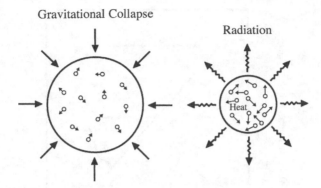

Figure 10.3 Gravitational collapse The collapse of an interstellar cloud of gas and dust (*left*), compresses the cloud, heats it and produces radiation (*right*). The arrows are pointed in the direction of motion of the gas atoms, and the lengths of the arrows denote their speed of motion.

As it turned out, a star begins it life on the main sequence. Once an interstellar cloud of gas and dust has become sufficiently massive, or after external compression has been provided, it must collapse to make stars, in much the same way that the Sun came into being. The mutual gravitational attraction of its parts will overcome the internal gas pressure and cause this cloud to start collapsing. As this protostar falls inward from all directions, the gas gains energy; a dropped stone similarly gains energy when it moves down to a pool of water. Some of the star's energy is converted into heat as the gas particles fall inward and collide with each other (Fig. 10.3).

When observations of young stellar clusters were combined with the theoretical studies of the Japanese astrophysicist Chiushiro Hayashi, the pre-main-sequence evolution of protostars of different masses was deciphered.[25] Their tracks in the H-R diagram initially move straight down, and subsequently turn to the left and continue that way until the protostar arrives on the main sequence (Fig. 10.4). The outward pressure of the hot gas, which is now heated by nuclear fusion reactions, prevents the star from collapsing further. It has settled down for a long, rather uneventful life as a main-sequence star, the longest stop in its life history.

So stars begin their lives on the main sequence and spend the majority of their time there. The giant stars belong to a subsequent and shorter-lived part of a star's evolution. As it turned out, stars do not move along the main sequence, but stay on it for many millions to billions of years.

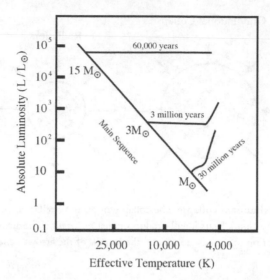

Figure 10.4   Protostars Evolutionary tracks of protostars of various mass in the Hertzsprung-Russell, or H-R, diagram, ending with their arrival on the main sequence when stars have begun burning hydrogen in their core. High-mass stars, which have greater luminosity than low-mass stars, are found at higher points on the main sequence and take a shorter time to arrive there. The luminosity, L, is in units of the solar luminosity, $L_\odot$, the mass in units of the Sun's mass, $M_\odot$, and the effective visible disk temperature is in units of degrees kelvin, designated K.

When all the hydrogen fuel has been used up in the core of a star, it can no longer support itself under the crush of gravity. It collapses inside, which increases the central temperature and density and opens up a new source of energy not previously available, while the surrounding stellar atmosphere expands to produce a red giant star.

The Way Stars Become Giants

It was the Estonian astronomer Ernst Öpik who in 1938 proposed how main-sequence stars might become giant stars.[26] He suggested that both the giant and main sequence stars shine by thermonuclear processes that follow a well-defined sequence of increasing core temperature. As a star evolves, the nuclear fuels burn from the center outward, and successively new nuclear fuels begin to burn at the center. For the giant stars, the exhaustion of the first processes begins earlier and the central temperatures rise to open up a new source of energy not available to main-sequence stars.

The course of the star's trajectory into the realm of the giants was mapped out in the H-R diagrams of globular star clusters, just as Strömgren had proposed, which brings us to Allan Sandage, who spent much of his life looking at the stars from the dark, cathedral-like domes of the largest telescopes.

Sandage was born in Iowa City, Iowa, on June 18, 1926, the only child of a Professor of Advertising and a homemaker mother. He began his undergraduate studies in 1943 at the Miami University in Oxford, Ohio, because his father was on the faculty there at the time, but two years later Allan was drafted into the Navy, where he spent 18 months training to be an electronics technician's mate. When he was discharged from military service, his father was moving to the University of Illinois, so Allan transferred there to finish his undergraduate degree in physics and mathematics.

In September 1948, he joined the first class of students to begin formal graduate studies in astronomy at the California Institute of Technology, as a self-described "hick who fell off the turnip truck." Allan became an observing assistant for Edwin Hubble, but since Hubble had no specific tasks suitable for a doctoral thesis, the German astronomer Walter Baade, who was working at the Mount Wilson Observatory, became Sandage's advisor and guided him in using the 60-inch (1.5-meter) telescope to detect faint stars in globular star clusters.

Sandage's investigations of the star cluster M 3 showed that its main sequence does not contain the hot, luminous O and B stars (Fig. 10.5)[27] This indicated that it has an age of at least 5 billion years, which was the time for these stars to consume all their core hydrogen fuel. That was roughly comparable to the geological and radioactive-dating estimates for the age of the Earth. [Later more precise estimates indicated an age of 11.4 billion years for M 3.]

When Hubble died in 1953, Sandage, a fresh Ph.D. of just 27 years old, inherited the task of using the 200-inch (5.08-meter) telescope on Palomar Mountain in California to measure the age and fate of the expanding Universe. After five years of careful measurements of the distances of galaxies, he obtained a new value for the rate of expansion of the Universe, which meant that the Universe is seven times older than was previously thought.[28] His measurements indicated an age of about 13 billion years if the Universe has been expanding at a constant rate, which is close to today's accepted value for the time since the Big Bang.

Sandage also continued to investigate the paths of stellar life by describing how the observed properties of both open and globular star clusters depend on their age. Stars within a star cluster are all of the same

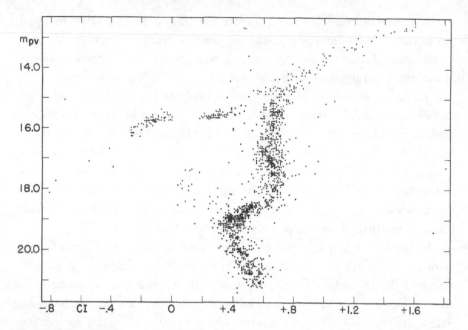

Figure 10.5   H-R diagram for M 3 The high mass stars in the globular star cluster M 3 have left the main sequence and evolved into the red giant branch (*top right*). Allan Sandage included this diagram in his 1953 doctoral thesis at the California Institute of Technology, entitled *A Study of the Globular Cluster M 3*. The vertical axis is the apparent photographic magnitude, m_{pv}, and the horizontal axis is the color index, CI.

approximate age, within a few million years of each other, dating back to the formation of the cluster many billions of years ago. They also began with the same initial composition of material, and exhibit a full range of stellar mass. And because the stars in a given cluster are all at the same distance from the Earth, we can obtain direct observations of their relative luminosity without knowing the distance.

All of the stars in a cluster begin shining when they arrive on the main sequence of the H-R diagram. As time goes on, the more luminous and massive stars evolve into the next phase of stellar life and the main sequence disappears from the top down. Very massive stars at the upper left of the main sequence become supergiants, and those with intermediate masses comparable to the Sun become red giants. A cluster H-R diagram can therefore be used as a clock, dating the age of the cluster and the stars in it by the place of their turnoff from the main sequence to become supergiants or giants. The lower the luminosity and temperature of the turnoff point, the older the star cluster is.

Sandage's investigations of stellar evolution in star clusters also involved theoretical calculations of just how long a main-sequence star's central fuel supply can last, and models of what happens when that fuel is used up. Martin Schwarzschild, the son of the German astronomer Karl Schwarzschild, was one of the first to examine these aspects of stellar evolution. After emigrating to the United States, Martin used theoretical models and primitive computers, developed by his Princeton University colleague John von Neumann, to chart the evolutionary trajectory of a star and compare it to the various kinks, bends, and gaps of missing stars in the observed H-R diagrams of globular star clusters.

Martin teamed up with Allan Sandage, whose observations included faint stars that connected the main sequence to the red giants.[29] They found that when the internal energy source is depleted in a main-sequence star of roughly a solar mass, the shrinking stellar core heats up and causes the star as a whole to swell up into a bloated red giant (Fig. 10.6). The internal energy

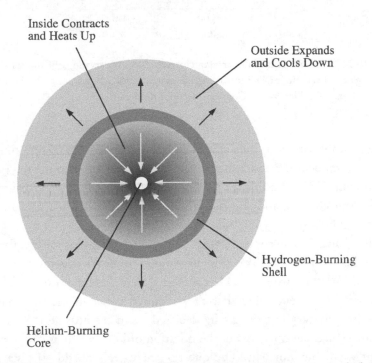

Figure 10.6   Formation of a giant star When a main-sequence star consumes the hydrogen in its core, the inside of the star contracts and heats up, while the outside expands and cools down. The center of this giant star eventually becomes hot enough to burn helium and stop the core collapse.

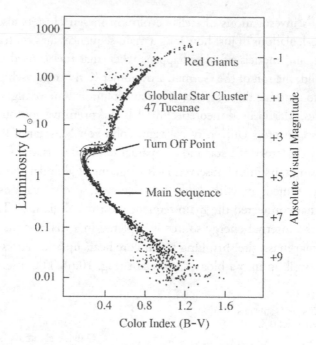

Figure 10.7   How old is a globular star cluster? These observations of the stars in the southern globular star cluster 47 Tucanae, also designated NGC 104, indicate it has an age of between 12 and 14 billion years. (Courtesy of James E. Hesser.)

released within the star is then spread over a much larger area, resulting in a lower disk temperature and a shift of the visible starlight into the red part of the spectrum. This accounts for the red giant branch in the H-R diagram of globular star clusters, and also permits determination of their ages that are between 10 billion and 14 billion years old (Fig. 10.7).

Allan Sandage also contemplated the mysterious reasons for life. By middle age, he had become plagued by two questions: "What is the purpose of life?" and "Why do we exist?" He concluded that these questions could not be answered by science, and that they required belief in the supernatural. "There is a mystery out there," he declared, "and it's outside the realm of science. Science can only answer the question of how, when and where, and perhaps what, but not why. The question of why is outside the scientific purview; but that's still part of the whole picture."[30] In this distinction between the how and the why, he was agreeing with Henry Norris Russell.

Sandage eventually became deeply spiritual in his outlook on life, the Universe, and the practice of astronomy, writing: "If there is no God, nothing makes sense.... And if there is a God, he must be true both to science and religion." The Big Bang, for example, might account for how the expanding Universe began, but "knowledge of the creation is not knowledge of the Creator, nor do any astronomical findings tell us why the event occurred. It is truly supernatural and by this definition a miracle."[31]

In order to explain the biggest why, and understand the mystery of existence, Allan Sandage became devoutly religious and appealed to something bigger, outside and beyond both observational astronomy and his own existence.

11. The Ways Stars Die

"And all about the cosmic sky,
The black that lies beyond our blue,
Dead stars innumerable lie,
And stars of red and angry hue
Not dead but doomed to die."

Julian Huxley (1933)[1]

The stars seem immutable, but they are not. They are all impermanent beacons that will eventually cease to shine. The exceptionally luminous stars, with the greatest mass, have brief lifetimes in astronomical terms, and will simply run out of energy in several million years. Other, intrinsically dimmer stars of lesser mass, settle down to rather uneventful lives lasting billions of years. But they will also inevitably perish, shining their substance away and returning to the darkness from which they came.

Dying stars do not disappear. They just change from one form to another. Their demise often results in the creation of a new star, like the phoenix arising from its ashes. This final resting state depends on the collapsing star's mass. Most stars, which have a mass comparable to the Sun's mass, end up as burned-out, Earth-sized, white dwarf stars. A more massive and luminous supergiant star can leave a city-sized neutron star behind, or be crushed into a stellar black hole.

The Winds of Death

Any star with a moderate mass, comparable to that of the Sun, will eventually balloon into a red giant star and shed its outer layers that are blown away. Astronomers watch these winds when they observe planetary

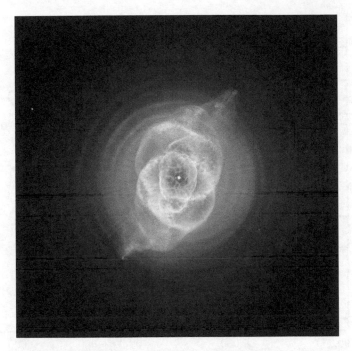

Figure 11.1   The Cat's Eye Nebula This planetary nebula designated NGC 6543, exhibits concentric rings, jets of high-speed gas, and shock-induced knots of gas. (A *Hubble Space Telescope* image courtesy of NASA/ESA/HEIC/the Hubble Heritage Team, STScI/AURA.)

nebulae (Fig. 11.1), which have round shapes and radiate bright emission lines.

When heated, a low-density gas will radiate these emission lines, so their presence indicates that a planetary nebula contains hot, rarefied gas. But at first, no one knew what the gas was.[2] The wavelength of one of the emission lines coincided with the lightest element hydrogen, but the chemical identification of two other emission lines remained a mystery for more than half a century. At first it was thought that a previously unknown substance dubbed *nebulium*, had been found. But the green nebular lines were eventually attributed to unusual states of known elements like oxygen and nitrogen, which have been synthesized within giant stars.[3]

The winds of death observed as planetary nebulae are therefore seeding interstellar space with these elements, along with the carbon and helium that have also been manufactured within giant stars. All of these heavy elements might be incorporated within future stars and their planets.

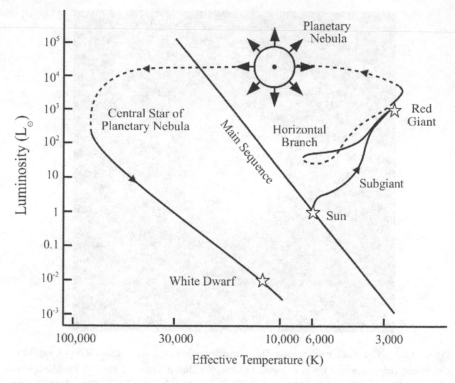

Figure 11.2   Formation of a planetary nebula and white dwarf star When a Sun-like star has used up its nuclear hydrogen fuel, it expands into a red giant star. After a relatively short time the giant star ejects its outer layers to form a planetary nebula, and its hot stellar core collapses to form an Earth-sized white dwarf star.

As a young planetary nebula is blown outward by powerful winds, it slowly grows in size, thins out and becomes transparent, revealing its source, the exposed core of a dying red giant that is collapsing inside (Fig. 11.2). Since the giant star has run out of fuel, there is nothing left to support it, and the core ends up as a true dwarf star. Such stars came to be called *white dwarf stars*, because of their initial white-hot color and small size. Their discovery was entirely unexpected.

Stars the Size of the Earth

The American astronomer Henry Norris Russell has recalled the discovery of the first hot, faint white dwarf star, which occurred during his visit in 1910 to the Harvard College Observatory. Russell thought it would be a good idea to obtain the spectra of certain stars, and Edward C. Pickering, director of the Observatory, asked for the name of one of these stars. Russell replied that the faint companion of Omicron Eridani, denoted o Eridani B, was an example. Pickering remarked: "Well, we make rather a specialty of being able to answer questions like that." And so he telephoned down to the office of Williamina Fleming and asked her to look up the star's spectral classification.

In about half an hour she reported that the spectrum of o Eridani B implied that it was a hot, white star. Russell was flabbergasted, and baffled about what it meant, for o Eridani B was a dim star of much lower luminosity than other hot, white stars. Then Pickering thought for a moment and with a kindly smile said: "I wouldn't worry. It's just these things which we can't explain that lead to advances in our knowledge."[4]

A few years later, in 1914, the American astronomer Walter S. Adams drew attention to the A0 spectral type of o Eridani B, which suggested a disk temperature of about 10,000 degrees kelvin, and noticed that it was surprising that such a hot star should exhibit such a very low luminosity. In the following year, Adams reported that the brightest star in the night sky, Sirius A, also has a faint companion that displays the spectral features of an intensely hot star.[5] What Adams did not point out explicitly was that the high disk temperatures in combination with the low luminosity meant that o Eridani B and Sirius B had to be very small — only about the size of the Earth.

These white dwarf stars are now known to be the inner, collapsed leftovers of dying red giant stars, exposed by the planetary nebulae that have carried off their outer atmospheres. Like a butterfly, a white dwarf star begins its observable life by casting off a cocoon that enclosed its former self.

Such stars can no longer ignite nuclear reactions, and their light must come from the slow leakage of the heat leftover from their former life inside a giant star. The resultant white dwarf star will slowly cool down and fade away, like a dying fire ember. Astronomers can measure their temperature to tell how long the white dwarf has existed, which is how crime detectives tell when a murder occurred, from the warmth of the corpse.

The rather ordinary stellar mass of a white dwarf has been compressed to a remarkably high mass density of up to a million times that of the Sun. At the time of their discovery, some astronomers thought that such a high density was impossible. The great English astronomer A. S. Eddington described the situation with:

> "We learn about the stars by receiving and interpreting the messages which their light brings to us. The message of the companion of Sirius when it was decoded ran: 'I am composed of material 3,000 times denser than anything you have ever come across; a ton of my material would be a little nugget that you could put in a matchbox.' What reply can one make to such a message? The reply which most of made in 1914 was — 'Shut up. Don't talk nonsense.'"[6]

Eddington nevertheless realized that there is nothing inherently absurd about the high mass density of white dwarf stars. Since all the electrons are stripped away from their atoms in the hot stellar interiors, the free electrons can be closely packed with the bare atomic nuclei, within the former space of the empty atoms. Within a decade, the new statistical laws of quantum theory were being developed, and they indicated that the densely packed electrons, rather than the nuclei, support a white dwarf star.[7]

When pushed together, the electrons resist being squeezed into each other's territory, darting away at high-speed just to keep their own space, somewhat like active dancers in a very crowded nightclub. This provides a pressure that resists further compaction and holds a white dwarf's immense gravitational forces at bay.

The small size and high mass density of the white dwarf stars have now been substantiated by measurements of their gravitational effect on the wavelengths of their spectral lines, and also confirmed by detection of their intense magnetic fields that have been amplified during the stellar collapse in which they were formed.[8]

In the very distant future, our Sun will become a hot, extended, giant star, and then shrivel up into a white dwarf star. This will be the final resting state for the vast majority of other stars, an insignificant cinder about the size of the present-day Earth.

When the Sun Dies

The Sun shines by consuming itself, and when its central hydrogen fuel is depleted the star's core will contract and heat up to burn helium. Its outer parts will then swell to gigantic proportions. Mercury will become little more than a memory, being pulled in and swallowed by the enlarged Sun. Our star will change its predominant color from yellow to red; dramatically increase its luminous output; boil the Earth's oceans away, and bake our once green planet into a dead and sterile place.

And if that doesn't wipe out living things, there will be no escape in the end. When its central helium is eventually used up, the Sun's fires will be forever extinguished. In a last gasp of activity, it will shed its outer layers of gas to produce an expanding planetary nebula, and the core of the once-powerful Sun will collapse into itself and squeeze its enormous mass into a white dwarf star that will eventually cool into darkness. There will be no possibility of life, as we know it, anywhere in its vicinity.

The English poet Lord Byron captured the essence of what the darkness might be like, writing at a time that the global ash of an active volcano, Mount Tambora, blocked out the light of the Sun, in a year without summer:

> "I had a dream, which was not all a dream.
> The bright Sun was extinguished … and the icy Earth
> Swung blind and blackening in the moonless air."[9]

A different fate awaits stars that are significantly more massive than the Sun.

Shrinking Way Down

A very massive star will not settle down as a white dwarf in its old age, but instead undergoes further collapse. The electrons will move faster and faster as a collapsing star gets smaller and smaller. And since you can't make an electron, or anything else, move faster than the speed of light, there is an upper limit to the electron's speed of motion and a maximum stable mass for a white dwarf star.[10]

While on board a ship taking him from Bombay to London in 1931, Subrahmanyan Chandrasekhar derived this upper limit at just 19 years of age, finding that a white dwarf star can be no more massive than 1.5 times the mass of the Sun.[11] In the same year, the Russian astrophysicist Lev Landau also found that very massive stars could not be supported against continued gravitational collapse at the endpoints of their life. As far as he could tell, they would just keep on collapsing to a point. Since extremely massive stars are now observed and do not show any such "ridiculous tendency," Landau concluded all stars heavier than 1.5 solar masses possess high-density "pathological regions" in which the laws of quantum mechanics are violated.[12]

When visiting the University of Cambridge in 1934-35, Chandrasekhar improved his calculations, and found that no white-dwarf equilibrium state is possible for a mass greater than about 1.46 times that of the Sun.[13] At the time, the English astronomer A. S. Eddington got into a bit of a row, as the English would say, with Chandrasekhar over the physical possibility of such a situation. At a meeting of the Royal Astronomical Society, A. S. stated: "Dr. Chandrasekhar had got this result before, but he has rubbed it in. ... I think that there should be a law of nature to prevent a star in behaving in this absurd way."[14] The gravity of the collapsing star would become strong enough to hold in its radiation so it couldn't be seen, and Eddington just did not like this idea.

A temporary way out of the impasse was found when it was realized that some stars explode when they die, and that their collapsing cores might be arrested by the formation of neutron stars.

Stars that Blow Up

As discussed in Chapter 6, there are two ways that stars can explode into a supernova at the end of their lives.[15] The *type I supernova* involves a white dwarf star that is a member of a close binary star system, with a companion that is a normal main-sequence star. If hydrogen flows from the expanding normal star onto its compact neighbor, it can push the white dwarf above its limiting mass and detonate an explosion. [Also see Chapter 6, Fig. 6.10).]

Another type of supernova, dubbed *type II*, happens to single, isolated stars without a nearby companion. When such a very massive star has used up all its nuclear fuel, it collapses and blows apart all by itself (Fig. 11.3). Nuclear reactions in the star continue, at ever-increasing central temperatures, until an iron core is produced. Since iron does not burn, no matter how hot the star's core becomes, the star can no longer support its own crushing weight and the iron core collapses, like a building with the foundation

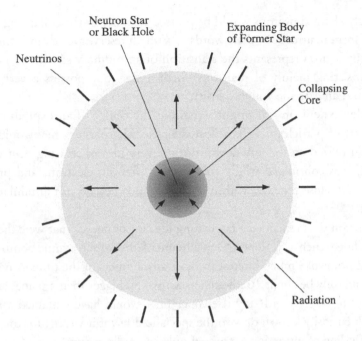

Figure 11.3 **Type II supernova** When an isolated star uses up all its nuclear fuel it blows up; its shattered remains are propelled into surrounding space; and its core is compressed by gravitational contraction into a neutron star or a black hole.

removed. When reaching mass densities approaching that of an atomic nucleus, the collapsing core bounces back and expels the outer parts of the star into space at supersonic speeds.[16] The dying star suddenly increases in brightness a hundred million times or even a billion fold, and in this kind of stellar explosion a neutron star or a black hole is left behind at the center.

Neither kind of explosion is forecast for the Sun's future. It will pass into its final resting state unaccompanied by a nearby stellar companion, and the Sun is nowhere near massive enough to explode by itself when it dies.

Dead Stars that Have Lost their Charge

Just two years after the discovery of the neutron, by James Chadwick in 1932, Walter Baade and Fritz Zwicky proposed that a neutron star might be left behind after a supernova explosion. The negatively charged electrons and positively charged protons would be pressed together in the collapsing stellar core to form neutrons. In their words: "With all reserve we advance the view that a super-nova represents the transition of an ordinary star into a *neutron star*, consisting mainly of neutrons. Such a star may possess a very small radius and an extremely high density."[17]

A dense neutron star might remain at the center of an exploding star, composed of particles that have lost their electric charge. They would have been neutralized on the way down. But the very idea of creating neutrons in this way was considered wildly speculative. After all, electrons and protons remain very close together within an atom, and they do not annihilate one another.

Neutron stars remained a fascinating idea for physicists that were theoretically inclined, such as J. Robert Oppenheimer father of the atomic bomb,[18] but they did not evoke much interest in most astronomers of the time. A neutron star would only be about 10 kilometers across, no bigger than a planet and far too small to be seen with any telescope. They would have remained a minor textbook curiosity known only to the specialized few if it weren't for the unexpected finding of mysterious, repeated pulses of radio radiation.

The Discovery of Pulsars

The long trail to the unanticipated discovery of radio pulsars began in England when Antony Hewish was sent to work with a top-secret wartime team, led by Martin Ryle, to produce electronic equipment for jamming the radar used by enemy night-fighter aircraft. At the end of the war in 1945, both Hewish and Ryle returned to the University of Cambridge where they pioneered different aspects of radio astronomy.

Ryle constructed a spread out array of modest-sized radio telescopes, connecting them to make several interferometer pairs and simulate a single large radio telescope. This improved the angular resolution of intense radio sources and the sensitivity needed to detect weaker ones.

Hewish completed his undergraduate degree in 1948 and received the Ph.D. degree four years later. By setting up two small antennas separated by about 1 kilometer and timing the variation of the radio intensity received at each site, he was able to measure the size and speed of structures in the ionosphere at high altitudes in the Earth's atmosphere that no one had observed before.[19]

After a decade of these investigations of the ionosphere, it was found, in 1964, that the observed intensity of some radio galaxies also changed when their radiation passed through the expanding solar atmosphere. When viewed though the wind-driven solar material, the observed radio waves sporadically blinked on and off in less than a second, in much the same way that stars twinkle when seen through the Earth's varying atmosphere. Such radio-intensity changes are called *interplanetary scintillations*, for they occur at large angular distances from the Sun and throughout the space between the planets.

The cosmic radio objects had to be of sufficiently small angular size to exhibit these scintillations, just as visible-light stars twinkle and the Moon does not because of its larger angular extent. And this meant that a study of the radio scintillations conveyed information about the size of the radio galaxies. Hewish and his colleagues therefore designed and built a radio telescope operating at a wavelength of 3.75 meters, or a frequency of 81 MHz, to study these effects. Together with his staff and graduate students, he constructed a huge array of 1,024 dipole antennas spread over an area of 4.5 acres and connected by miles of wire to a radio receiver and a paper chart recorder that sampled the signals every 0.1 seconds, the time scale of the scintillations. [Grazing sheep were used to cut the grass beneath the array.]

In July 1967, the research group began a sky survey, making repeated observations in order to observe the interplanetary scintillations over a wide range of angular distances from the Sun. Analysis of the chart recordings was the project of graduate research student Jocelyn Bell, who was assigned the task of identifying the positions of all scintillating sources in the sky. This was to eliminate terrestrial radio interference that would not reappear at the same position.

When examining the long flow of paper sent out from the recorder, Bell found a bit of "scruff" whose unchanging position in the sky was not associated with any known cosmic radio source. The intensity variations were also observed when the array was pointed away from the Sun. The effects of the Sun's wind should have been small, and any cosmic radio source could not be scintillating.

Hewish asked Jocelyn to look at the unexpected, fluctuating signals with a high-speed recorder to find out what might be causing them. That led to the even more astonishing detection of a succession of short radio pulses repeating at regular intervals of just over one second, or to be precise with a repetition period of 1.3372795 seconds. The first radio pulsar had been discovered.[20] Absolutely no one had foreseen the existence of the quick, rhythmical cosmic radio signals. As often happens in astronomy, an unanticipated discovery had been made while looking for something else, or as Tony Hewish stated: "It was rather like miner's looking for tin and unexpectedly finding gold."

By the time the discovery was ready for publication, in 1968, evidence of other radio pulsars was found in the existing chart recordings, and within three weeks a second paper announced the discovery of three additional radio pulsars.[21] This triggered searches by other radio astronomers for additional previously unknown pulsars with large radio telescopes using rapid time sampling, rather than the long signal integration times formerly used. In less than a year, the list of pulsars had been expanded to over two dozen.

Tony Hewish was a devout Christian, of the Anglican faith, but his first religious experience did not occur in a church. It happened on a golf course, where he had a supernatural, mysterious "numinous experience" and felt the presence of a benevolent power behind the entire Universe.

He believes there is a great deal of mystery in both science and religion. Physicists, for example, believe in unseen virtual particles that come in and out of existence much too rapidly to be detected, and according to Hewish this "helps you to get in the right frame of mind to realize that religious mysteries can exist, and be reasonable without defeating common sense." For Hewish, the Christian belief in God was vital to his considerations of what it means to be human and the purpose of our existence.[22]

Jocelyn Bell was brought up as a Quaker, attended a Quaker boarding school, and never abandoned the quiet Quaker belief in an inner "Light." Like Hewish, she has long had an awe-inspiring, wondrous sense of the presence of divinity. She also continued with spiritual growth and prayer and communion with God throughout her life.

It was the stillness and beauty of the natural world that helped lead Jocelyn to a sense of reverence, gratitude and joy, to transcendent "moments of eternity." When discussing the implications of astronomy for our beliefs, she wrote that although we live in a physical Universe that is mostly dark and largely empty, there is "a loving, caring, supportive, enpowering God, a God who works through people."[23]

The Nobel Prize in Physics for 1974 was awarded jointly to Sir Martin Ryle and Antony Hewish, in particular to Ryle for the aperture synthesis technique and Hewish for his decisive role in the discovery of pulsars. Fred Hoyle and other people criticized the Swedish Nobel Committee for not sharing the award with Jocelyn Bell who played a pivotal role in the discovery of the first pulsar, but she never was upset over the exclusion.

This brings us to the discovery that the radio pulsars are rotating neutron stars.

Radio Pulsars are Rotating Neutron Stars

The Austrian-born American astronomer Thomas Gold proposed that the pulsed radio emission is produced by a rapidly rotating neutron star with an intense magnetic field.[24] He assumed that the radiation is emitted in a beam, like a lighthouse, oriented along the magnetic axis (Fig. 11.4). An observer sees a pulse of radio radiation each time the rotating beam flicks across the Earth. And because the neutron star's beam could be oriented at any angle, the beams of many pulsars would miss the Earth and a lot of them would therefore remain forever unseen.

Like any good scientist, Gold suggested definitive observational tests of his ideas. He realized that a neutron star could rotate very fast, due to the conservation of spinning motion at the time of its formation from the collapse of a larger, slowly rotating star. Gold therefore predicted that astronomers would detect pulsars with shorter periods than those first discovered, which was confirmed by the discovery of the Crab Nebula radio pulsar that spins around 30 times every second.[25] It is located at the position of the very star thought to be the neutron star remnant of the Crab Nebula supernova whose brightening was documented by the Chinese in 1054 (Fig. 11.5).

Tommy Gold also noticed that a spinning neutron star will gradually lose its rotational energy and slow down, and successfully predicted that this would cause a slow lengthening of the radio pulsar periods with time. The loss of rotational energy inferred from the observed period increase of the Crab Nebula pulsar is exactly that needed to keep the nebula shining at the present rate for about 1,000 years, ever since the supernova explosion that was associated with the pulsar's birth.

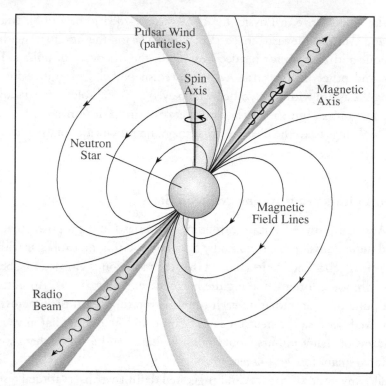

Figure 11.4 Radio pulsar Electrons encircle the powerful magnetic field of a spinning neutron star and emit intense, narrow beams of radio radiation that can sweep across the Earth as the star rotates. A bright pulse of radio emission, called a pulsar, can be observed once every rotation of the neutron star.

Once the appropriate technology was developed, a different sort of x-ray pulsar was unexpectedly discovered in close orbit about a perfectly normal star.

X-ray Pulsars

Unlike cosmic visible light or radio radiation, cosmic x-rays do not reach the ground. They are totally absorbed in the Earth's atmosphere, and have to be observed using detectors lofted above it.

The Sun was the first known source of cosmic x-rays. In the 1950s, the United States military wanted to find out why the Sun's activity occasionally

Figure 11.5 The Crab Nebula supernova remnant The optically visible light of the Crab Nebula, designated as M 1, displays expanding filaments and an inner amorphous region. It is the remnant of a supernova explosion observed nearly 1,000 years ago, in the year 1054. The south westernmost (*bottom right*) of the two central stars is a radio pulsar and neutron star that is spinning 30 times a second. (Courtesy of NASA/ESA/J. Hester and A. Loll, ASU.)

disrupted their radio communications, and astronomers at the Naval Research Laboratory looked into the matter. Herbert Friedman, Richard Tousey and their naval colleagues used instruments aboard captured German V-2 rockets, and subsequently their own sounding rockets, to show that solar x-rays of varying intensity alter the Earth's outer atmosphere, the ionosphere, which was being used to reflect radio waves used in global communications.[26] The apparently serene Sun, an unchanging disk of brilliant light to our eyes, has no permanent features in x-rays, which describe a volatile, unseen world of perpetual change.[27]

If the Sun was any guide, then x-rays could not be observed from any other star. Even the nearest stars other than the Sun would be too far away, and their x-ray emission too faint to be detected with existing instruments. But there might be unknown sources of cosmic x-ray radiation, and Riccardo Giacconi's group at the American Science and Engineering Company designed the sensitive equipment needed to search for them. Their pioneering rocket flight in 1962 was successful in detecting the first stellar x-ray source, which has extraordinary and unforeseen properties.[28] Its x-ray luminosity is a thousand times its visible light intensity, and a thousand times the entire luminosity of the Sun at all wavelengths.

The exciting potential of scanning the sky in x-rays was supported by NASA, which funded two satellites dedicated to x-ray astronomy. The first one was launched on December 12, 1970 from the Italian site of San Marco, off the Coast of Kenya. Since this date coincided with the seventh anniversary of the independence of Kenya, the satellite was given the name *Uhuru*, the Swahili word for freedom. The second dedicated x-ray telescope is known as the *Chandra X-ray Observatory*; it was launched by NASA on July 23, 1999 and was named after the theoretical astrophysicist Subramanyan Chandrasekhar.

The *Uhuru* observations revealed that some x-ray sources are regularly emitting a succession of pulses with periods of seconds, like radio pulsars except in x-rays.[29] After analyzing a year of observations of one of the x-ray pulsars, Centaurus X-3, the *Uhuru* scientists found a longer and regular pattern of intensity changes, increasing and decreasing in strength every 2.1 days as the result of an orbiting companion star. An additional year of scrutiny revealed that the period of this x-ray pulsar was getting shorter, which meant that its rotation was speeding up and not slowing down like radio pulsars.

The spinning neutron star is fed by a spillover from a nearby ordinary star (Fig. 11.6). The in-falling gas swirls and spirals around and down into the neutron star, like soapsuds circulating down into a bathtub drain. Friction between the rapidly moving inner parts of the whirling disk and its slower-moving outer parts heats the gas to millions of degrees kelvin, emitting luminous x-rays. When the in-falling material lands it gives the neutron star a sideways kick, increasing its rotational energy, speeding it up, and causing the rotation period to become shorter as time goes on.

Although most radio pulsars are alone in space without a nearby companion, some have been found in close embrace with another neutron star. The first such discovery resulted from new techniques used to provide accurate timing measurements of known radio pulsars, and to search for faint unknown ones. This binary radio pulsar had much greater significance than anyone had foreseen.

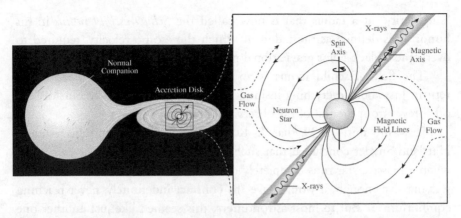

Figure 11.6 X-ray pulsar The outer atmosphere of an ordinary star, detected in optically visible light, spills onto its companion, an invisible neutron star. The flow of gas is diverted by the powerful magnetic fields of the neutron star, which channel the in-falling material into the magnetic poles and produce x-ray emission that sweeps across the sky as the neutron star rotates.

Disappearing from Sight into a Black Hole

In 1784 the Reverend John Michell, an English clergyman and natural philosopher, suggested that a star might be so massive, and its gravitational pull so powerful, that light could not escape it. As he wrote: "All light from such a body would be made to return to it by its own proper gravity."[30] The star would be invisible.

The French astronomer and mathematician, Pierre-Simon de Laplace popularized the idea in the late 18th century, and subsequently showed that light could never move fast enough to escape the immense gravitational attraction of some compact stars.[31] Their matter might be so concentrated, and the pull of gravity so great, that light could not emerge from them, making these stars forever dark and imperceptible. This unseen star is now known as a *stellar black hole*.

On the theoretical front, there has been decades of frenetic activity in describing the nuances of stellar black holes, without much regard for the observable Universe. It all began when the German astronomer Karl Schwarzschild derived the solutions to Einstein's *General Theory of Relativity* outside a point mass, while serving as an artillery officer on the Russian front during World War I (1914–1918).[32] He showed that it contains a "singularity" at a radius that is now called the *Schwarzschild radius* in his honor. It is defined as the radius at which the escape velocity, required to overcome a black hole's gravitational pull, is equal to the speed of light.

The Schwarzschild radius is an ultimate boundary, the place of no return. Just about everything inside it is disconnected from the outside, cut off forever from the rest of the observable Universe.

It wasn't until 1939 that J. Robert Oppenheimer and his student Hartland Snyder concluded that such a stellar black hole might be created when a massive star runs completely out of thermonuclear fuel. Provided it was massive enough, it would have "to contract indefinitely, never reaching equilibrium."[33] But to most astronomers, this seemed like just another one of those fanciful theoretical speculations, and in a few years Oppenheimer was off to other interests at Los Alamos, New Mexico, where he directed both the theoretical and experimental studies that led to the explosion of the first atomic bomb.

From an astronomer's point of view, the difficulty is that there is no way you can see such a black hole all by itself. It is black because no light can leave it, and it is a hole because nothing that falls into a black hole can escape. It does not absorb, emit, or reflect radiation. And since it could not be seen no astronomer could tell if it was even there.

With remarkable foresight, Michell also speculated, in 1784, that such an unseen dark star might nevertheless betray its presence by its gravitational effects on a nearby, luminous star in orbit around it.

In modern extensions of this idea, a black hole is detected indirectly when the outer atmosphere of a nearby visible star spills over into the immense gravitational influence of the black hole. This material swirls around and down into the black hole, orbiting faster and faster as it gets closer — as the result of the ever-increasing gravitational forces.

The in-falling particles collide with each other as they are compressed to fit into the hole, heating the material to temperatures of millions of degrees kelvin. At these temperatures, the gas emits almost all of its radiation at x-ray wavelengths. So, the way to find a stellar black hole is to look for two stars that are in close orbit, one a normal visible star and the other unseen except for its x-rays.

The archetype of a stellar black hole is Cygnus X-1, located in the constellation Cygnus. It was one of the first x-ray sources to be discovered, and was known for its rapid fluctuations in x-ray intensity.[34] Cygnus X-1 is accompanied by a bright, blue supergiant star whose periodically shifting spectral lines indicate that it is revolving about an invisible companion of more than 8 times the mass of the Sun.[35] Any normal star with this mass would be very bright and easily seen through a telescope, but it emits no detectable visible light. That massive, dark companion is a stellar black hole, and many of them have now been identified in this way.

Exceptionally impressive black holes inhabit the centers of galaxies. They are massive, scaled-up versions of stellar black holes, with millions if not billions of times the mass of the Sun packed into a region just a few light-years across. Like stellar black holes, the supermassive ones cannot be directly observed. Their presence is inferred from the orbital motion of nearby visible stars whose trajectories are guided by the otherwise invisible black holes.[36] Without the gravitational pull equivalent to millions and even billions of Sun-like stars, the fast-moving stars would fly away from the unseen centers of the galaxies.

These supermassive black holes were proposed as the power source for the intense radio emission of radio galaxies and quasars.[37] The huge black holes are apparently consuming more stars than they can fully digest, and continuously ejecting high-speed, radio-emitting electrons in opposite directions along the rotation axis of a black hole. [Also see Chapter 6, Fig. 6.9.]

By measuring the sharp rise in orbital velocities of stars at close distances from galaxy centers, astronomers have weighed unseen supermassive black holes in nearby galaxies. One of them is located at the center of our Milky Way Galaxy. Observations of the orbital motions of adjacent stars imply the mass of the central black hole is 4 million times the mass of the Sun — that is, 4 million solar masses not shining but rather gravitationally confining the observed stellar orbits (Fig. 11.7).[38]

Since it feeds on surrounding gas, dust and stars, you might wonder why the supermassive black hole at the center of our Galaxy has not consumed the entire Milky Way. Fortunately, its reach is limited, and does not extend across such enormous distances.

The astronomer's black-hole concept has entered everyday language as a common metaphor, as a place where something might become forever lost, like childhood, a former love, or memory. A black hole might also remind people of their fears of being consumed, as in a job or marriage, or even destroyed.[39]

There is no known force that can overcome the powerful gravitational pull of a black hole. It is a one-way street, a path of no return, where you can

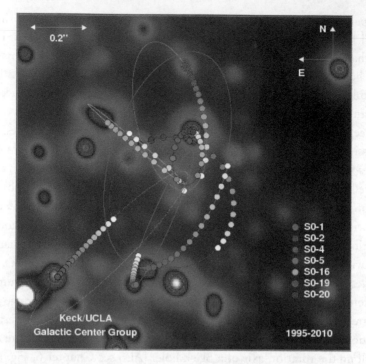

Figure 11.7 Super-massive black hole centered in our Galaxy The orbits of infrared stars near the center of our Milky Way indicate that a super-massive black hole is located there; it has a mass of 4.1 million times the mass of the Sun. The angular scale is denoted at the top left, designating 0.2 seconds of arc or 0.2″. (Courtesy of Andrea M. Ghez/UCLA galactic center group.)

go in but can't come out. It resembles human death, when someone might be buried underground within a dark, silent grave. Once you have crossed that line, you can't come back. You are gone, lost, consigned to oblivion forever. For a star, only its gravitation remains.

This brings us to dark interstellar places that stars arrive from and disappear back into.

12. Darkness Made Visible

"Now entertain conjecture of a time
When creeping murmur and the poring dark
Fills the wide vessel of the Universe."

William Shakespeare (1599)[1]

"God made darkness his secret place."[2]

Night Falls

As the Earth rotates, the daytime side of our planet turns away from the Sun's light and becomes night. The darkened continents lose their borrowed sunlight; a quiet hush passes over the land; and time seems to hesitate as the Sun drops below the horizon. Cool breezes refresh the air, fish come up to feed, swallows swoop through the air, and moonflowers open to the nightly glow.

Most familiar things slide away when evening falls. Daytime boundaries are lost, our familiar surroundings lose their form, houses become wrapped in a blanket of darkness, and entire cities seem to drain themselves of life. They dissolve into the shadows, fuse into the enfolding night, and vanish from view.

The night can bring delight.[3] Fireflies, tiny specks of living light, can rise to twinkle in the dark, and the Moon and stars come into view. But even the stars seem insignificant and lost, as they try to light up the immense blackness that separates them. They resemble tiny fires in a cold, silent world.

Many astronomers nevertheless revel in the dark quiet of the night and the splendor it brings into view. There are luminous stars that heat and illuminate adjacent regions (Fig. 12.1), and immense, mysterious black places that are far bigger than the glowing regions (Fig. 12.2).[4] At first sight, they look empty and without substance, characterized by an absence of anything

Figure 12.1   Rosette Nebula Hot O and B stars in the core of this nebula exert pressure on the nearby interstellar material, trigger star formation, and heat the surrounding gas to a temperature of about 6 million degrees kelvin. (Courtesy of KPNO/CTIA.)

we know, but closer scrutiny reveals that all that darkness is not an emptiness. It contains numerous fine, solid particles of dust that absorb and scatter starlight.[5]

The ancient Chinese sage Lao-tzu wrote: "Darkness within darkness, the gate to all understanding,"[6] and he was right. Astronomers know there is always something in the dark. It is often the kernel and substance of things to come. In the Earth, subterranean seeds lie slumbering below the cold winter surface. They are waiting to rise, bloom, and unfurl their beauty in

Figure 12.2   Dark clouds New stars may be born in the dense molecular clouds of the Carina Nebula. Energetic stellar winds and intense radiation from nearby massive stars are sculpting its outer edges. (A composite *Hubble Space Telescope* image courtesy of NASA/ESA/ Hubble Heritage Project/STScI/AURA.)

the warmth of the spring. Humans begin their lives in the darkness of their mother's womb, and stars form within immense dark places.

Flowers, humans, and stars; they all keep appearing out of the dark and disappearing back into it. It may be the same darkness from which all of us and everything else have come and to where we are all going.

There has to be something more than dust out there if entire stars are being spawned from it. Just as the stars are themselves mainly composed of the lightest element hydrogen, it is hydrogen atoms that provide the main substance of interstellar space. This material has been additionally enriched with lesser amounts of heavier elements manufactured inside former stars.

When it ceases to shine, an entire star can be returned to the darkness from which it came. The celebrated American poet T. S. Eliot wrote some appropriate lines, when he was probably thinking of humans rather than stars:

> "O dark dark dark. They all go into the dark,
> The vacant interstellar spaces, the vacant into the vacant."[7]

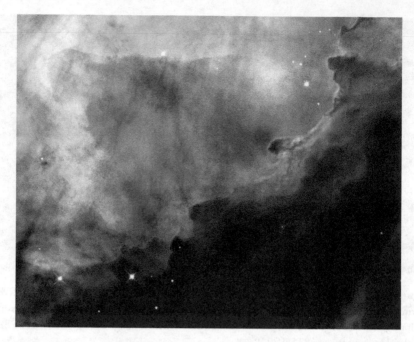

Figure 12.3   Light and dark Intense ultraviolet radiation from young, massive stars illuminates and shapes interstellar gas within the Omega Nebula, which is also designated as M 17. (A *Hubble Space Telescope* image, courtesy of NASA/ESA/J. Hester, ASU.)

To glimpse the full beauty of the world, we want to focus on both the light and the dark. It is the combination of brightness and shadow that make sunrise, sunset, or a stormy day so captivating. You see the contrast in the shadows cast on the ground by a drifting cloud, or in the Milky Way, where: "The unseen dark plays on his flute, and the rhythm of light eddies into stars and Sun, into thoughts and dreams."[8] Light and dark are always there, coexisting and depending on one another (Fig. 12.3).

Detecting the Unseen

For centuries astronomers viewed the Universe through the visible light rays emitted by stars. Their scrutiny began with unaided eyes, and was then extended using telescopes whose lenses and mirrors gathered in more light. This enabled the resolution of details that could not be seen before, and the detection of faint, otherwise invisible objects, such as the galaxies.

In the 20th century, unique telescopes, new technology, and novel detection equipment enabled astronomers to penetrate the darkness with radio waves and x-rays. They widened our cosmic vision by detecting invisible worlds that had been hidden in the dark for millennia. It was found that much of the Universe resides in the darkness, and remains out of direct sight to human eyes even with the aid of a visible-light telescope. Most of these unseen cosmic objects were totally unexpected, and no one predicted or even imagined many of them.[9]

What you observe anywhere in the Universe depends on how you look at it. The cold, dark interstellar spaces, for example, emit most strongly at long radio wavelengths, and very hot cosmic gases, with temperatures of millions of degrees kelvin, shine brightly at short x-ray wavelengths. It is only stars like the Sun, with disk temperatures of several thousand degrees kelvin that radiate intense visible light. It is all a matter of perspective, with invisible radiation disclosing some things, and visible light revealing others.

Cosmic Radio Broadcasts

It wasn't until the closing years of the 19th century that anyone knew a thing about radio waves. The German physicist Heinrich Hertz first generated them in his laboratory in 1886,[10] and the Italian entrepreneur Guglielmo Marconi pioneered global radio communications as a commercial venture shortly thereafter.[11] But these developments did not lead directly to new windows on the Universe, and so had little bearing for astronomy.

Radio emission from the Milky Way was inadvertently discovered in the 1930s when the Bell Telephone Laboratories assigned Karl Jansky the task of tracking down and identifying natural sources of radio noise that were interfering with transatlantic radio transmissions at a wavelength of 14.6 meters. He constructed a rotating antenna, which pointed sideways at the horizon and permitted the identification of the interference, including the intermittent radio static produced by lightning discharges from distant thunderstorms.

Fortunately, the antenna's wide field of view also pointed part way up into the sky, and thereby detected an extraterrestrial hiss of unknown origin, which was comparable in intensity to terrestrial lightning. By observing the variation of its intensity as a function of direction and time of arrival, Jansky established that the radio emission had to originate from outside the Solar

System, and that the most intense radiation was coming from the direction of the constellation Sagittarius and the center of our Galaxy.[12]

This serendipitous discovery of cosmic radio broadcasts was reported by newspapers throughout the world, including the *New York Times* whose front-page headline for May 5, 1933 read: "New Radio Waves Traced to the Center of the Milky Way." Jansky even appeared on a radio program, which rebroadcast his "star noise" so listeners could hear "the hiss of the Universe" by a direct long-line connection from the Bell Labs field station in Holmdel, New Jersey to a New York Broadcasting studio.

Although Jansky wanted to follow up his incredible discovery, the Bell Labs assigned him to other engineering problems more directly related to their objectives. The country was in the throes of the Great Depression, when jobs were scarce and Jansky was most likely glad to be employed. He never returned to research in astronomy after 1935, and died in 1950 at the age of 44 without ever receiving any scientific award for his profound discovery.

Moreover, astronomers almost completely ignored the result. Jansky did not publish in an astronomical journal, and his radio techniques were so much outside the conventional methods of astronomy that no traditional observatory contributed to new knowledge about it. So astronomers did not become aware of the different way of looking at the Cosmos until the 1940s, when an electrical engineer, amateur astronomer, and avid ham radio operator, Grote Reber, confirmed and extended Jansky's findings and published them in the *Astrophysical Journal*.[13] With the occasional help of a local blacksmith, he built with his own hands a 9.6-meter (31-foot) metal, dish-shaped radio telescope in the backyard of his home in Wheaton, Illinois, and after a few years of trying he used it to detect radio noise coming from the Milky Way.

Most stars other than the Sun do not emit detectable radio waves, and intense cosmic radio radiation originates in the space between them. This emission is attributed to cosmic-ray electrons traveling within interstellar space at a speed close to that of light. They are similar to the cosmic-ray protons and electrons impinging on the Earth's atmosphere,[14] but only the electrons emit radio waves.

Although long radio waves reveal the presence of invisible high-speed electrons in interstellar space, one particular radio wave, just 21 centimeters long, discloses cold hydrogen atoms in the space between the stars.

The Fullness of Space

Since most stars are mainly composed of hydrogen, it was supposed that the space they arose from ought to contain large amounts of the substance. And because new stars are still forming today, there ought to be plenty of hydrogen atoms out there right now. These atoms move slowly at the freezing temperatures of interstellar space and gently knock against each other, which stimulates invisible radio emission. The prediction and discovery of this radiation follows a paper trail of international scope, remarkable coincidences, and a professional courtesy of bygone times. It is also a tale of two graduate students and their ever so smart advisors.

This story began near the end of World War II (1939–1945), when astronomers at the Leiden Observatory in the Netherlands obtained smuggled copies of the *Astrophysical Journal* from America. When the Dutch astronomer Jan Oort read about Grote Reber's unexpected discovery of cosmic radio static, he asked his graduate student Hendrik C. "Henk" Van de Hulst to find out if there were any spectral lines in the radio spectrum. He investigated the matter, and predicted a radio wavelength spectral line that might be detected from interstellar regions of electrically neutral, or unionized, hydrogen atoms.

Van de Hulst realized that these sources, now designated *H I regions* and pronounced H one regions, would be very cold and that most of the atoms would be in their lowest energy state. In this condition, the lone electron of the hydrogen atom has two possibilities in the direction of its spin, or rotation, and a rare collision between two of the atoms could cause one of them to flip over and reverse its spin direction. The atom is then in an unstable configuration, so its electron will soon flip back to its original state. This releases a small amount of energy and produces radiation at a wavelength of 21 centimeters. As Van de Hulst pointed out, these spin transitions will occur rarely in the cold tenuous interstellar gas; a given atom only undergoes the spin flip once every 11 million years. But an observer might well detect them when looking through the vast extent of interstellar space.

The prediction was published in 1945 within an obscure Dutch journal *Nederlands tijdschrift voor natuurkunde*, with an obtuse article title of "Radio Waves from Space: Origin of Radio Waves,"[15] which did not tell the reader very much about what Van de Hulst had found. The Soviet theorist Iosif

S. Shklovskii then confirmed the prediction seven years later, with greater detail, but in the Russian language.[16]

At about this time, Harold I. "Doc" Ewen, a graduate student at Harvard University, became interested in radio astronomy. Ewen's advisor, Edward M. Purcell, asked his wife, Beth, to translate Shklovskii's paper about the 21-centimeter radiation, and after reading it Purcell encouraged Ewen to build a radio receiver to search for it. They also had a copy of Van de Hulst's original work, translated from the Dutch.

Ewen's previous experience made him especially suited for building the radio detector. After completing undergraduate study at Amherst College, he joined the Navy, where he was trained in radar, first at Princeton University and then at the Radiation Laboratory of the Massachusetts Institute of Technology. As a radar officer in a naval airborne squadron during World War II (1939–1945) he learned all about repairing radio equipment, sometimes without any spare parts.

After the war, Ewen entered Harvard University on the GI bill, and to make ends meet he worked at the cyclotron particle accelerator of the University's nuclear laboratory. His interest in detecting hydrogen became "a weekend thing, strictly a hobby on the side" of his real job. He put together a "detecting machine" using electronic components that were scavenged from other places, or brought with a $500 grant and $300 from Purcell's pocket. Much of the equipment was borrowed every Friday, and returned each Monday, from the nuclear laboratory using a wheelbarrow.

Accurate measurements of the wavelength of the expected hydrogen emission, using terrestrial atomic hydrogen in a laboratory at Columbia University, enabled Ewen to tune his receiver to precisely 21.106 centimeters. And at Purcell's suggestion, the receiver was switched between the wavelength of the expected signal and an adjacent one, with a difference that might contain the anticipated hydrogen radiation. [Such a wavelength-switched, or frequency-switched, receiver has now been widely adapted by radio astronomers to remove unwanted noise from an observed signal containing a spectral feature.]

The receiver was connected to a simple horn antenna that was built of plywood, lined with copper sheeting, and mounted on a ledge just outside a high window. It pointed in a fixed direction up into the sky, which swept by as the Earth rotated, and the 21-centimeter transition was detected when the Milky Way entered the antenna beam.

Hendrik Van de Hulst happened to be serving as a Visiting Professor at the Harvard College Observatory at the time, and the Australian astronomer Frank Kerr was visiting Harvard on a Fulbright grant. So Ewen and Purcell had them over to describe their discovery, and urged them to have their people confirm the result.

At the meeting, the Harvard team learned for the first time that the Dutch group at Leiden had been actively trying to detect the radio transition for several years. So a description of the wavelength-switched receiver was provided to Van de Hulst, leading to the conversion of the Dutch system, and in a gracious move, which would be unheard of in today's competitive scientific world, Purcell insisted that publication of their discovery be held up until the Dutch group confirmed it.

Even though the Australians were not actively pursuing a search for the hydrogen radio emission, and did not have a detection system in operation at the relevant wavelength, they assembled the necessary components soon after receiving word from Kerr, and repeated the detection about a month after the Dutch had. A coordinated report from all three centers was then published in the journal *Nature* in July 1951.[17, 18]

The detected line was seen in emission, and the line profiles could be used to estimate the temperature of the dark interstellar hydrogen atoms. Using this technique, the Leiden research group found that they are very cold indeed, with a temperature of about 100 degrees kelvin, well below the freezing temperature of water at 273 kelvin.

After the pioneering detection, the Navy called Ewen to return to active duty because of the Korean conflict, and the Harvard astronomy department did not immediately follow up his discovery. It was the Dutch and Australian groups that took advantage of the fact that the 21-centimeter radiation propagates right through the curtains of interstellar dust that hide most of our Galaxy from view at the optical wavelengths of visible light.

By laboriously turning an old German radar antenna in various directions by two small hand cranks every few minutes for much of a two-year period, the Leiden group collected observations to show that the interstellar gas seen in the northern hemisphere describes four extended arm-like features. In the meantime, the Sydney group used a radio telescope that looked straight up, and could not be moved, to discern similar features in the southern Milky Way as the Earth's rotation turned these regions past the telescope's field of view. The two surveys were combined to give the Leiden-Sydney map of our Galaxy in which neutral hydrogen atoms are concentrated in several elongated, nearly circular features that resemble spiral arms. (Fig. 12.4).[19]

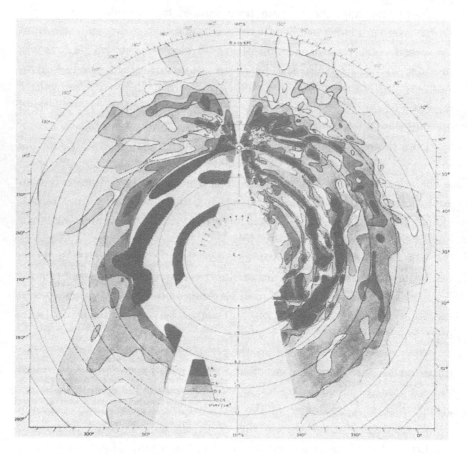

Figure 12.4 Arms of the Milky Way The distribution of the 21-cm radiation from hydrogen atoms in the plane of our Galaxy, the Milky Way, displays arm-like features. Large circles are spaced 6,500 light-years apart and the small circle and the letter S in the top middle denote the Sun. [Adapted from Jan H. Oort, Frank J. Kerr, and Gart Westerhout: "The Galactic System as a Spiral Nebula," *Monthly Notices of the Royal Astronomical Society* **118**, 379–389 (1958).]

Soon after the discovery of interstellar atomic hydrogen at radio wavelengths, astronomers began to speculate about the possibility of detecting molecules with radio waves, which are sensitive to the coldest clouds of interstellar matter. But observations of any molecule first required accurate measurements of the frequencies or wavelengths of its spectral features in the terrestrial laboratory. Charles Townes played an important role in these pioneering investigations.

Masers, Lasers, and Interstellar Molecules

Charles H. Townes grew up in a devoted Southern Baptist family. He spent his childhood on a 20-acre cotton farm in Greeenville, South Carolina, where he attended public schools and then graduated from the local Furman University, a Baptist Institution, in 1935 at the age of 19. He completed work for a Masters degree in physics at Duke Univeristy a year later, and then entered the graduate school at the California Institue of Technology where he attended lectures by Albert Einstein, Willie Fowler, Robert Millikan, J. Robert Oppenheimer, Richard Tolman, and Fritz Zwicky. In Townes' account of these and later times, he notes how chance conversations and encounters of life lead, in totally unpredictable ways, to the events that shape a career.[20]

In 1939, with a fresh Ph.D. degree in hand, Townes went to work for the Bell Telephone Laboratories in New York, where he helped design radar bombing systems for the U.S. Air Force throughout World War II (1939–1945). The radar was supposed to operate at a short wavelength of 1.25 centimeters, which would provide high angular resolution, but it was eventually scrapped because radiation at this wavelength cannot travel very far in the Earth's atmosphere before it is absorbed by water molecules there.

This got Townes interested in how molecules absorb and emit energy, and he began a program of molecular spectroscopy at microwave wavelengths, from 0.1 centimeters to 10 centimeters, that he continued for two decades as a faculty member at Columbia University. During this time, he and graduate students built a maser filled with ammonia gas, NH_3, in which the molecule turns itself inside out, like an umbrella in a high wind. [The nitrogen atom, N, vibrates back and forth in the plane of the three hydrogen atoms, H, at the same 1.25-centimeter wavelength he had investigated for radar purposes.]

Townes has recalled that the former and current chairmen of his department in Columbia University, who were both Nobel laureates for their work with atomic and molecular beams, asked him to discontinue this work, since they didn't think it would work and their research depended on the same source of support as his. But Townes kept on with what interested him, and was thankful that he had come to Columbia with tenure, so they couldn't fire him for doing what he wanted. He was also grateful that his funding agencies in the military did not have any specific expectations or instructions. Townes was instead thrilled, intrigued and stimulated by the beauty of nature, from a calm sea to a stormy one, from an atom to a field of wild flowers, or an insect, bird, fish, star or galaxy.[21]

The term *maser* is an acronym for microwave amplification by stimulated emission of radiation; the reader may be more familiar with the *laser*, where the letter "*l*" denotes visible light.[22] When focused to a small point, laser beams can produce intensities of light billions of times that at the Sun's surface.

The basic idea had been anticipated by Einstein in 1917 when he explained how radiation could induce, or stimulate, still more radiation when it hits an atom or a molecule provided the energy of the incoming photon equals the energy that can be lost by an atom or a molecule in making a transition from a higher to a lower energy state.

Charles Townes led a somewhat nomadic life, moving from place to place and entering other fields, to turn over new stones and see what is under them. While at Columbia, he took an interest in astronomy and at the request of Hendrik van de Hulst attended an international symposium on radio astronomy, in 1955, where he specified what molecules might be found in interstellar space. This included his precise laboratory measurements of the vibrational and rotational transitions of molecules that might be detected at microwave or radio wavelengths, including the carbon monoxide, CO, water, H_2O, and hydroxyl, OH, molecules.[23] But radio astronomers could not observe these molecules in interstellar space until new methods of spectral analysis and radio telescopes with surfaces accurate to a few centimeters were constructed.

Townes served as Vice President and Director of Research at the Institute for Defense Analysis in Washington, D.C. from 1959 to 1961, and during the next six years, from 1961 to 1967, he was Provost at the Massachusetts Institute of Technology, or MIT for short. In 1963 Alan Barrett used a new digital receiver designed by Sander Weinreb at MIT to obtain the first observations of intersellar OH at a wavelength of 18.005 centimters, or a frequency of 1,667 MHz, that Townes had specified.[24] The discovery was confirmed within days by other radio astronomers, with an unexpected consequence.

In some places, the OH molecules were emitting energy of high brightness temperature at the spectral line wavelength, rather than absorbing it. They were acting like cosmic masers that were being pumped into high energy levels by radiation from stars and by collisions. It was subsequently found that interstellar water molecules are even more powerful masers, which can emit much more power at a single wavelength than the Sun does at all wavelengths. It is an amazing coincidence that the masers invented in the terrestrial laboratory were also operating in interstellar space.

In 1964 Townes received the Nobel Prize in Physics, which he shared with Nicolai G. Basov and Alexander "Sasha" Prokhorov of the Lebedev Physical Institute in Moscow, for "fundamental work in the field of quantum electronics, which has led to the construction of oscillators and amplifiers based on the maser-laser principle." In simpler terms, they received the prize for their invention of the maser and laser.

Throughout his career, Townes has reflected on the similar goals, faith and insights of science and religion. Scientists have faith that there is an unchanging order in the Universe, from the Earth to the larger Cosmos, and that this order is understandable by human beings who seek to discover it. Religious persons have faith that they can understand the purpose and meaning of the Universe and how we fit into it; their discoveries often come about by great revelations. Townes states that discovery in science does not usually come from the "scientific method" of logical deduction from observed facts, and that these discoveries are often accidental, intuitive, and sudden. He believes that science and religion will ultimately converge and join together in a common pursuit of the truth.[25]

Townes was appointed University Professor at the University of California in 1967, with a location at the Berkeley campus. In this position, he and his graduate students discovered ammonia and water in interstellar space.[26] This was soon followed by the detection of the embalming fluid formaldehyde, and carbon monoxide by other groups.[27]

These discoveries triggered an avalanche of molecular searches in which groups of young radio astronomers armed with the latest laboratory measurements engaged in an extraordinarily competitive fight to be the first to detect the next interstellar molecule. The net result has been the discovery of a pharmaceutical array of hundreds of interstellar molecules, including complex organic molecules such as ethyl alcohol, or ethanol, the substance that gives beer, wine and liquor their intoxicating power.

The molecules reside within dense, dark and dusty, interstellar clouds that typically span up to 120 light-years and harbor a total mass of up to a million times the mass of the Sun, mainly in the form of molecular hydrogen. These hydrogen molecules do not emit radio spectral lines, but their presence can be inferred by their collisional excitation of carbon monoxide.

It is the tiny, solid dust particles in the giant molecular clouds that help block out the harsh radiation in space, and enable chemical reactions to form complex, delicate molecules from the atomic constituents of the interstellar gas. These clouds are black and exceedingly cold, with temperatures of only about 10 kelvin, radiating almost exclusively in the microwave region of the electromagnetic spectrum.

Under the right circumstances, a giant, massive molecular cloud collapses under its own weight, eventually forming up to a million stars. In fact, some stars are now forming in these clouds. They are the present-day incubators of newborn stars.

In the 1980s, it was additionally discovered that atoms and molecules are not the only substance out there in the black cosmic places. The distant parts of our Galaxy are rotating so fast that they ought to fly apart unless the gravitational force of some other, unseen material is holding it all together. The existence of this dark matter was also anticipated by observations of motions within clusters of galaxies.

Dark Matter

The Universe is permeated with an unusual substance that is not of the ordinary kind we can see with our eyes. It is known as *dark matter*, which sounds like the title of a mystery thriller or an old-fashioned *film noir*. Astronomers use the name for something that emits no visible light or any other kind of radiation, and hence is dark, but it interacts gravitationally like ordinary matter. It is something we cannot directly see and yet know must be there.

The astronomer Agnes Clerke noticed the possible cosmic implications of this dark matter more than a century ago, when the entire Universe was thought to consist only of stars. She wrote of dark stars, whose presence might be inferred from their gravitational effects on the motions of visible stars, and stated that: "Unseen bodies may, for ought we can tell, predominate in mass over the sum total of those that shine. They supply possibly the chief part of the motive power of the Universe."[28]

How do we know that dark matter exists when we can't see it? Astronomers infer its presence by observing the way visible stars, interstellar gas, or galaxies move. The gravitational forces of the dark matter hold swirling stars and gas in at the edges of galaxies, and keeps clusters of galaxies from flying apart. It is something like noticing a powerful wind from the way it shakes a tree.

Back in 1937, the eccentric Fritz Zwicky, a Swiss astronomer working at the California Institute of Technology, showed that the Coma cluster of galaxies ought to be breaking apart unless large amounts of unseen matter were keeping it intact.[29] Zwicky estimated the mass of individual galaxies in the cluster from their luminosities and from their observed rotational motions, and he used the random motions of the galaxies in the cluster to estimate the mass of the entire cluster required to hold it together. He found that the sum of the masses of the galaxies was not enough to keep the cluster dynamically stable. That is, there was not enough luminous matter in the visible galaxies to gravitationally hold them together and keep the cluster intact. Their motions had to be balanced by the gravitational pull of some hidden substance that was much more massive than the sum of the masses of the visible galaxies the cluster contains.

A few years earlier Zwicky had even introduced the term *dunkle materie*, or "dark matter" for the invisible stuff, and concluded: "If this is confirmed, we would arrive at the astonishing conclusion that dark matter is present with a greater density than luminous matter."[30]

As Zwicky proposed, the dark matter also acts like a zoom lens, and magnifies remote galaxies too faint to be otherwise seen.[31] When the light from a distant galaxy passes through an intervening cluster of galaxies, the light rays are bent, diverted, focused and magnified by its unseen dark matter. Observations of this gravitational lensing have been used to detect the presence of dark matter in a cluster of galaxies, and to determine the concentration and distribution of the unseen matter within the cluster (Fig. 12.5).[32, 33]

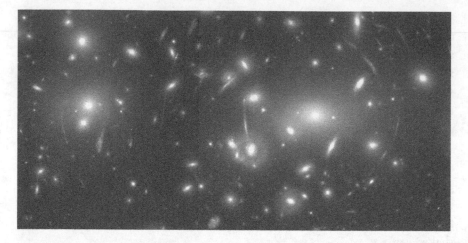

Figure 12.5 **Cluster of galaxies and gravitational lens** A *Hubble Space Telescope* image of a rich cluster of galaxies, designated Abell 2218, whose invisible dark matter deflects light rays and magnifies, brightens and distorts images of objects that lie far beyond the cluster. (Courtesy of NASA.)

Dark matter is not solely confined to clusters of galaxies. Substantial imperceptible matter envelops individual galaxies as well. This totally unexpected result was discovered by measuring the rotational motions of gas and stars in the outermost reaches of nearby spiral galaxies. Such movements are inferred from the Doppler shifts of spectral lines of bright stars and emission nebulae, seen at optical wavelengths, or interstellar hydrogen gas detected at radio wavelengths near 21 centimeters.

Astronomers naturally thought that the bright center of a spiral galaxy was the most massive region, which gravitationally controlled the motions of the stars outside it. In this case, the stars and interstellar material would revolve around the massive center at speeds that decrease with their distance from it, just as the more distant planets move with slower speeds around the central massive Sun. But the outer peripheries of some nearby spiral galaxies, which can be scrutinized in detail, are moving too fast to be constrained by anything that can be directly seen. They must be held in by the gravitational attraction of large amounts of invisible dark matter.

Around 1970, Vera C. Rubin and W. Kent Ford, at the Carnegie Institution of Washington, D.C., showed that at least one galaxy, the nearby Andromeda Nebula, or M 31, was not rotating as expected.[34] When they measured the rotational velocities from the visible-light spectra of emission-line regions, they remained rotating at high speeds at large distances from the center of the galaxy.

Radio astronomers then showed that remote clouds of hydrogen atoms are spinning about the centers of other spiral galaxies unexpectedly fast. The orbital velocity of the gas remains constant with increasing distance from the center of spiral galaxies; well beyond the visible stars (Fig. 12.6).[35] The radio astronomical results were not fully appreciated until Rubin, Ford and their colleagues turned their full attention to the spinning movements of a host of spiral galaxies in the following decade.[36] Like Andromeda, their outermost regions circle the centers of these galaxies as quickly as the inner parts closer to the centers. Substantial amounts of unseen, non-luminous matter are required to keep the outer parts from moving out of their galaxy's control.

This indicated that most of the material within galaxies is not concentrated near their center, where the visible-light luminosity is greatest. Their outer parts are filled with an invisible, non-luminous substance, a dark matter, which is far more massive than anything we can see. It also meant that the spiral galaxies don't end where their light does, and that they extend further than the eye can see even with the aid of any telescope.

Well outside the boundary of its visible light, there are appreciable amounts of unseen matter that keeps the fast-spinning visible material connected to a galaxy and holds it together. Astronomers estimate that each nearby spiral galaxy is immersed within, and enveloped by, a halo of dark matter at least 10 times more massive than its visible component, extending as far as a million light-years.

There is more than meets the eye in our Galaxy as well. The mass of its visible inner regions, inferred from the orbital motion of the Sun about the galactic center, is about 100 billion Suns. But the rapid motions of dwarf satellite galaxies, which revolve about our Milky Way at distances of up to a million light-years, indicate that a great reservoir of unseen matter also envelops our Galaxy. In order to hold onto and retain these dwarf companions, the invisible parts of our Galaxy must outweigh the visible ones by a factor of about 10. Most of this mass does not lie within the plane of the Milky Way, but beyond it in a dark halo, which extends out to a million light-years from the galactic center.

The visible stars of nearby spiral galaxies must be sandwiched within a similar massive darkness to retain their shapes.[37] About 90 percent of the mass of galaxies has to reside in this much larger unseen halo giving off

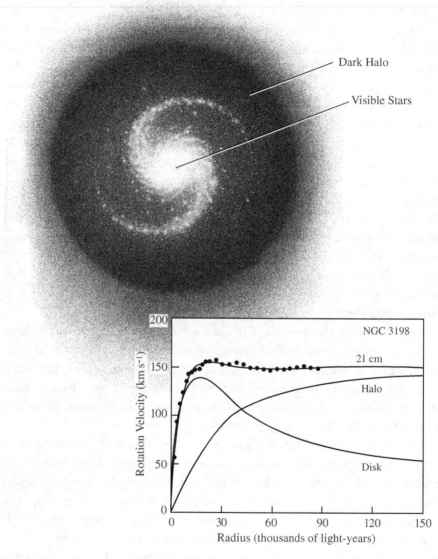

Figure 12.6 Dark halo envelops a spiral galaxy The rotation velocity of the spinning spiral galaxy NGC 3198 plotted as a function of distance from its center (*bottom*). The observed emission from the 21-cm line of hydrogen atoms indicates that a halo of dark matter contributes most of the mass at distant regions from the center (*top*). The rotation velocity is given in units of kilometers per second, or km s^{-1}. [Adapted from T. S. Van Albada and colleagues, "Distribution of Dark Matter in the Spiral Galaxy NGC 3198," *Astrophysical Journal* **295**, 305 (1985)].

neither visible light nor any other radiation to let us know it is there or what it might be composed of.

These discoveries were a consequence of the pioneering investigations of Fritz Zwicky and Vera Rubin, who both preferred to find out things by themselves.

Working Alone — Fritz Zwicky and Vera Rubin

After graduate study in mathematics and physics at the Swiss Federal Institute of Technology in Zurich, Fritz Zwicky travelled to the California Institute of Technology, abbreviated Caltech, where he began investigating cosmic rays and eventually became Professor of Astronomy.

Fritz preferred to work alone, and once planned to write an autobiography entitled *Operation Lone Wolf*. Eccentric, aggressive and independent, he liked to be right, and to show others were wrong. [Also see Exploding Stars in Chapter 6] Zwicky thought his distinguished colleagues stole his ideas and hid their own mistakes. He also rarely gave a grade in his courses better than an average C, since in his opinion the students never did very well.

Fritz was indeed quite a clever and accomplished fellow. He pioneered the study of extragalactic exploding stars, the supernovae, foretold the existence of city-sized neutron stars, inferred the presence of dark matter in clusters of galaxies, predicted the existence of cosmic gravitational lenses, and contributed to research on cosmic rays, jet propulsion, and the first man-made satellites.

Vera Rubin also liked to work mainly alone, but she preferred to avoid controversy and was never as outspoken as Zwicky. As a child Vera liked to watch the stars move through the sky every night, and by high school she knew she wanted to become an astronomer. Vera Cooper had graduated from Vassar College and acquired a husband, Robert Rubin, before she was 20. She joined him in graduate studies at Cornell University, where her 1951 masters degree speculated that galaxies might move around on their own with an extra sideways motion apart from their outward expansion. Such peculiar, streaming motions were subsequently observed.

When Vera's husband started a job at the Johns Hopkins Applied Physics Laboratory, Vera entered a doctoral program in astronomy at the nearby Georgetown University, where George Gamow served as her advisor. Her doctoral thesis, completed in 1954, involved the clumped distribution of galaxies. By that time she had two children, and another two by 1960. Her astronomical career slowed down as a result of family life, and did not accelerate until she got a job in 1965 at the Department of Terrestrial Magnetism, abbreviated DTM, at the Carnegie Institution of Washington. It was known for the freedom it gave its researchers without overt publication pressures.

At the DTM, Vera Rubin joined W. Kent Ford who had built an image tube device that enabled them to record spectra of galaxies in two to three hours rather than the thirty or more hours that it then took with other methods. In the early 1970s Vera began using these spectra to measure the rotation velocities of spiral galaxies at various distances from their centers, partly because of plain old curiosity about how that motion might be related to the galaxy's structure, and also because no one else was doing it. This led to the discovery that spiral nebulae, including our own Galaxy, are embedded within extended halos of dark matter.

Vera disliked mediocrity, and insisted on working on problems outside the main stream of astronomy. As she expressed it: "I liked doing things that other people were not doing, [tended] to work pretty much alone," and "did not interact with theorists at all."[38] She saw no conflict between her Jewish religion and science, and viewed them as separate aspects of her life.

Zwicky's critical attitude extended to religion, for he thought God was not needed to account for the already miraculous wonders of Nature.

The discoveries of extensive amounts of invisible dark matter in the Universe were subsequently confirmed and extended by spacecraft observations of the cosmic background radiation.

Elusive Dark Matter and Enigmatic Dark Energy

Two kinds of dark matter have been found to exist. The first kind is made of the familiar baryons that make up ordinary terrestrial material, as well as stars and galaxies. The name of these baryons is derived from the Greek *barys* for "heavy," and the protons and neutrons that provide most of the weight of atoms are baryonic. For lack of a better name, the other kind of dark matter has been named *non-baryonic matter*. This substance had to be around to give the first galaxies shape and form.

As the expanding Universe cooled, some regions of matter must have gathered together by their mutual gravitation to become the first stars and galaxies. However, the expansion of the Universe would pull any primeval baryons apart as soon as they started to coalesce. The extra gravitational attraction of dark matter is required to clump primordial fluctuations of the early Universe into the clusters of luminous galaxies that are now found in the Universe.

The first few minutes of the Big Bang provide clues to the amount of baryons in visible and invisible form, both then and now. The hydrogen, deuterium, and most of the helium that are now observed in the Universe originated back at these early times. When the observed amounts of these light elements are compared with the calculated quantities produced in the Big Bang, it is found that there has to be roughly 10 times as many invisible baryons in the Universe as visible ones that make up the stars. So the bulk of the baryons now residing in the Cosmos are not luminous, at least in visible light.[39]

Even these unseen baryons cannot fully account for dark matter. Detailed analysis of the irregularities in the cosmic background radiation, by instruments aboard *WMAP* and *Planck*, indicate that 82% of the total mass of the Universe does not consist of ordinary atoms or their sub-atomic consitutents.[40, 41] In other words, most of the dark matter is unlike any matter we know about, and the amount of this unknown substance far transcends ordinary baryonic matter in visible or invisible form. So perhaps it might also be called *dark gravity*, since its gravitational attraction is the only thing we know for certain about it.

This unseen substance is not the only unsolved mystery. There is the dark energy inferred from the runaway expansion of the Universe. The *Planck* all-sky survey indicates the Universe contains about 5% ordinary matter, 26% dark matter, and 69% dark energy. Astronomers are now actively seeking to know more about dark matter and dark energy.

13. Primordial Light

"In the beginning God created
the Heaven and the Earth.

And the Earth was without form,
and void; and darkness was upon
the face of the deep. And the Spirit
of God moved upon the face of the waters.

And God said, Let there be light:
and there was light."

The Holy Bible, Genesis[1]

Soldier, Priest and Cosmologist

Georges Henri Joseph Édouard Lemaître was born on July 17, 1894 in the Belgian industrial town of Charleroi, the eldest son in a deeply religious, Catholic family. After attending the Sacred Heart High School in Charleroi and the College Saint Michel Preparatory School in Brussels, he entered the College of Engineering in Louvain, graduating in civil engineering in 1913. Lemaître began training as a mining engineer, but World War I (1914–1918) intervened and changed the course of his life.

The German army invaded Brussels on August 4, 1914, and Georges volunteered as a soldier in the Belgian army five days later. He spent the entire war exposed to its horrors, which included his participation in bloody house-to-house fighting and witness to the first poison gas (chlorine) attack in the history of warfare. Lemaître received the Croix de Guerre for his wartime bravery.

At the close of the war, he obtained a master's degree in mathematics and physics at the Université Catholique de Louvain, while also studying Aristotle and Aquinas. In October 1920 he entered the Belgian Maison Saint Rombaut as a seminarian, where he fulfilled ecclesiastical studies for the Roman Catholic priesthood and maintained his mathematical interests.

Lemaître had won a scholarship from the Belgian government to study abroad, and two weeks after being ordained, in September 1923, he left for the University of Cambridge in England to work with A. S. Eddington. The following year, in 1924, he traveled to the other Cambridge, in the United States, where he studied at the Massachusetts Institute of Technology, abbreviated MIT, and interacted with Harlow Shapley at the nearby Harvard College Observatory.

While in the United States, the Belgian cleric toured the country, and met with both V. M. Slipher at the Lowell Observatory in Flagstaff, Arizona and Edwin Hubble at the Mount Wilson Observatory in California. When Lemaître returned to Belgium, he therefore knew all about Slipher's observations of the redshifts, or outward velocities, of spiral nebulae and Hubble's determinations of a few of their distances, which Lemaître explained by proposing the expansion of the Universe from a primeval explosion.

A Day without Yesterday

In 1917 Albert Einstein tried to apply his *General Theory of Relativity* to a spatially finite and eternal Universe without any motion.[2] But there was a problem that was even obvious from Newton's simpler theory of gravitation. The so-called fixed stars cannot stay at rest with respect to one another. The unrelenting gravitational attraction between individual stationary, or non-moving, stars would eventually pull them together so the entire stellar system would collapse. Einstein therefore introduced an anti-gravity repulsion term, called the *cosmological constant*, to keep that from happening.

The extra term represented a universal repulsive force of an unknown and undetected form of energy that permeates space and exerts an outward pressure that opposes gravity. In contrast to other physical forces, which decrease in strength with increasing distance, the cosmological repulsion was assumed to be weak over short distances and stronger at large ones. Einstein adjusted the value of this cosmological constant so that its anti-gravity force would exactly counterbalance the combined gravitational attraction of all the known stellar matter in the Universe, keeping it from collapsing and making it stay put.

At A. S. Eddington's request, the Dutch astronomer Willem de Sitter had been submitting descriptions of the astronomical consequences of Einstein's theory for publication in the *Monthly Notices of the Royal Astronomical Society*. His third paper on the subject, published in 1917, included an empty Universe that would permit movement without a cosmological constant.[3]

It was time to put a little matter in de Sitter's moving world or a little motion in Einstein's non-moving, material one, which was stressed by A. S. Eddington at the January 10, 1930 meeting of England's Royal Astronomical Society. When news of the meeting reached Georges Lemaître, at the Université Catholique de Louvain, he wrote Eddington to remind his former teacher that he had solved the problem with an expanding model, which he had given Eddington three years earlier.

With remarkable foresight, the Belgian priest and astronomer had applied Einstein's theory to observations of spiral nebulae, and described them by a Universe that is in a state of expansion, with all the spirals moving away from one another at high speed.[4] He even derived the now-famous Hubble law in which the recession velocities of galaxies increase in proportion to their distance, two years before Hubble observed it. Others had previously used the complex theory to invent imaginary worlds, but Abbé Lemaître was the first to use it to explain the one and only true Universe that had already been observed by astronomers.

Eddington corrected his embarrassing oversight of Lemaître's work with a letter to the journal *Nature* that drew attention to the work, and by sponsoring an English translation of Lemaître's 1927 paper entitled "A Homogeneous Universe of Constant Mass and Increasing Radius accounting for the Radial Velocity of Extra-Galactic Nebulae."[5] In this explanation, space is swelling, ballooning outward, and carrying the extragalactic nebulae, or galaxies, with it, somewhat like birds riding on an otherwise unseen wind.[6]

Almost no one noticed Lemaître's prophetic 1927 publication. It initially had little impact on astronomers, and was only acclaimed years later when it was realized that his work applied to the real observed Universe. Another reason for the neglect was that his paper was written in French in the obscure *Annales de la Société scientifique de Bruxelles*. The Belgian cleric also tended to keep his ideas to himself, and did not present his findings to the general public.

Lemaître naturally wondered what the Universe has expanded from. If it is now expanding and growing larger, then a backward extrapolation implies a time when the Universe was incredibly small and in a state of high compression. Near the beginning of the expansion, he imagined, the galaxies must have been squeezed into a uniform mass of very small size, as tiny as a single atom or even the nucleus of that atom. Lemaître even supposed that all of the mass of the Universe once existed in the form of a unique atom and quantum of energy.

As he explained it in 1931:

"The present state of quantum theory suggests a beginning of the world very different from the present order of Nature." ... "We [then] find all the energy of the Universe packed in a few or even a unique quantum ... [It was] in the form of a unique atom, the atomic weight of which is the total mass of the Universe. This highly unstable atom would divide in smaller and smaller atoms by a kind of super-radioactive process."[7]

This primeval atom and single, undifferentiated quantum of energy divided and subdivided into a larger and larger number of energy quanta through a process of radioactive decay. According to Lemaître, such a beginning coincided with the origin of all things, including mass, space and time.

By this time, Edwin Hubble had published, in 1929, his velocity-distance relation for spiral nebulae, and Lemaître's paper was heralded as a brilliant synthesis of theory and observation. As he later explained, Lemaître was in part led to this conclusion by the similar billion-year lifetimes of long-lived radioactive elements, such as uranium and thorium, and the expansion age of the Universe.

As the result of his imaginative interpretations of the expanding Universe, Abbé Lemaître became something of a celebrity in the 1930s, lauded by the press and scientists alike. On May 19, 1931 the *New York Times*, for example, reproduced almost his entire 1931 *Nature* article on "The Beginning of the World From the Point of View of Quantum Theory."

Lemaître developed his concepts into a widely read book entitled *The Primeval Atom: An Essay on Cosmogony*, first published in French in 1946 and in English translation six years later. In this interpretation, the observable Universe began when a cold, super-dense primeval atom started to disintegrate. It resulted in the acceleration of cosmic rays, and to the eventual formation of stars and galaxies.

In this marvelously poetic and imaginative scenario, the beginning of everything occurred on a day without a yesterday, at the origin of time when the expansion of the Universe began. Since that time:

> "The [subsequent] evolution of the world can be compared to a display of fireworks that has just ended: some few red wisps, ashes and smoke. Standing on a well-chilled cinder, we see the slow fading of the suns, and we try to recall the vanished brilliance of the origin of the worlds."[8]

Reviving the Cosmological Constant

There was one problem with the new theory for the expanding Universe. The age of the Universe looked at first sight to be younger than geological estimates for the age of the Earth. When using Hubble's 1929 value for the expansion rate of the Universe, the beginning of the expansion was pegged at 1.8 billion years ago. Something wasn't quite right, for radioactive elements had been used to clock the oldest rocks on the Earth's surface at more than twice that age.

Neither Abbé Lemaître nor A. S. Eddington were at all troubled by this age discrepancy, which could be resolved by the delayed outward thrust of a cosmological constant. As we have seen, this anti-gravity repulsion had been initially proposed by Einstein in 1917 to keep the Universe in a static, non-moving condition. After it was realized that the Universe is expanding, Einstein abandoned the cosmological constant, and stated in 1931 that the *ad hoc* term was "greatly detrimental to the formal beauty of the theory." He instead collaborated with the Dutch astronomer Willem de Sitter to propose an expanding Universe described by the equations of his relativity theory without the anti-gravity term.[9]

Both Eddington and Lemaître nevertheless remained convinced that a non-zero cosmological constant was essential to an expanding Universe. To them, it was a real physical force that permeates the entire Cosmos. They supposed that the Universe had no definite origin in time and might have been resting in a pre-existing, immobile state before its expansion began.

Since the repulsion force of the cosmological constant increases with distance, it can overcome the gravitational attraction that holds the Universe together, and a slow expansion will then begin. Once the outward thrust gained the upper hand, the Universe would start flying apart and gradually thin out. This would lessen the overall gravitational attraction, and make the Universe less able to resist the cosmic repulsion, propelling it to faster and faster speeds as time went on.

Although Eddington and Lemaître agreed on the importance of this force, they had different views about the beginning of the Universe. Lemaître proposed that it began during the disintegration of his hypothetical primeval atom, but Eddington disliked the notion of an explosive beginning, which he found distasteful, and wrote: "As a scientist I simply do not believe that the present order of things started off with a bang … Philosophically, the notion of a beginning of the present order of Nature is repugnant to me," and he subsequently asserted that: "We cannot give scientific reasons why the world should have been created one way rather than another. But I suppose that we all have an aesthetic feeling on the matter … It has seemed to me that the most satisfactory theory would be one which made the beginning *not too unaesthetically abrupt.*"[10]

Eddington then continued his disdain with: "If you prefer the view (favored by Lemaître) that the Universe started with the thunder of an explosion, there is nothing in our present knowledge to gainsay you; only it seems inartistic to give a Universe, built to contain a natural cause of expansion, an additional shove off at the start."[11] That built-in natural cause was the cosmological constant.

It was all a matter of timing. Either the expansion started at its explosive beginning in the distant past, or else the expansion was a peripheral event unrelated to the origin of the Universe, which might even have existed for an eternity. The unsettled controversy was continued in the mid-20th century in two opposing worldviews.

A Universe that Had no Beginning

Two young astronomers from Austria, Hermann Bondi and Thomas Gold, and separately the English astronomer Fred Hoyle, proposed that the expanding Universe had no specific beginning in time and might last forever. Bondi and Gold had traveled to study at the University of Cambridge in the years immediately preceding World War II (1939–1945), but did not meet until May 1940 when the English government sent them into internment in Canada as enemy aliens. They were released by the end of 1941 and sent to join the English astronomer Fred Hoyle in work on the English naval radar systems at the Admiralty Signals Establishment. After the war, in 1945, Bondi and Hoyle returned to Cambridge, while Gold stayed with naval research until 1947, when he also came back to Cambridge and joined Hermann Bondi in proposing a novel Steady State theory for the expanding Universe.

Bondi and Gold were very critical of the application of Einstein's *General Theory of Relativity* to the Universe at large. As they pointed out, you could adjust the cosmological constant to accommodate almost any related observation, and the theory thereby lost its simplicity and uniqueness. In their view, any speculative theory that could never be tested by definitive observations was downright unscientific.

They therefore proposed a Universe that had no beginning, and supposed the Cosmos has always existed, presenting an unchanging Steady State on the largest scales of space and time.[12] The thinning of matter caused by the continuous expansion of the Universe is, in their view, compensated for by the perpetual creation of new matter in the increasingly empty spaces of the Universe.

The Steady State hypothesis would be consistent with the supposed universality of the laws of physics, which were assumed to be applicable everywhere and at all times throughout the Universe. They proposed a perfect cosmological principle in which the Universe is unchanging in both space and time, at least when the largest possible scales are concerned.

The new hypothesis acknowledged the inescapable fact that the galaxies are moving apart, but it supposed that they have always been doing so in an eternal Universe without beginning or end. As the Universe expands, newly formed galaxies continuously fill the intergalactic voids produced when older galaxies move away from one another. The dispersal and creation of matter therefore balance each other forever in an eternal Steady State, with the expansion taking things apart as fast as new matter comes into being. If just one hydrogen atom were created in every cubic meter of space per year, on average, it would be sufficient to keep the overall Universe unchanged with time.

This creation of new matter out of the nothingness of space might seem preposterous, but to Bondi and Gold it was no harder to accept than the supposed creation of matter all at once at the beginning of time.

Fred Hoyle had a different approach to the Steady State. He thought you should get the equations first and used Einstein's *General Theory of Relativity* to obtain them, without making use of the cosmological constant.[13] By introducing a new term in the same way that Einstein had introduced the cosmological constant, Hoyle was able to obtain an automatic balance between the expansion of the Universe and the origin of new matter.

At this time, and probably decades before that, most astronomers had nevertheless become tired of speculations based upon Einstein's mathematically complex theory. The behavior of galaxies and the eventual fate of their expansion could be described perfectly well by Newton's simpler gravitational theory without any space-time curvature or a cosmological constant. The astronomers needed some definitive observations that could be applied to the Universe at large, and they eventually got it in 1965, when the remnant radiation of the Big Bang was discovered.

A Hot, Radiant Beginning

It was one thing to describe the current expansion of the Universe, and quite another to specify how the expansion began. Abbé Lemaître realized the dilemma, suggesting an explanation with:

> "It remains to find the cause of the expansion. We have seen that the pressure of the radiation does work during the expansion. This seems to imply that the expansion has been set up by the radiation itself."[14]

But what caused the radiation in the first place?

While working at the Johns Hopkins University in Maryland in the late 1940's, George Gamow and his colleagues proposed that the observable Universe resulted from the detonation of a cosmic bomb. They were mainly interested in where the chemical elements came from, and suggested that they were cooked in the hot exploding oven of the infant Universe.

Instead of focusing on the mathematics of the early Universe, as Lemaître had before him, Gamow was interested in its physical properties. He found that conditions were relatively simple and uncomplicated in the distant past, when compared to those of today's Universe. Since gases get hotter when they are compressed and cool when they expand, Gamow concluded that the

Universe had to be incredibly hot in its earliest, most compact state, with a temperature of about 10 billion, or 10^{10}, degrees kelvin. It was so exceptionally hot in the beginning of the expansion that radiation was indeed the most powerful thing around. As Gamow put it: "One may almost quote the *Biblical* statement: 'In the beginning there was light,' and plenty of it."[15]

A single temperature would characterize the radiation, and that temperature would slowly drop as the expansion of the Universe progressed. In 1948 Ralph A. Alpher and Robert C. Herman predicted the present value of this temperature in a short paper intended to correct some mistakes in Gamow's account of the evolution of the Universe. An ending sentence, inserted almost as an afterthought, stated that the radiation should still be around, cooled to about 5 degrees above absolute zero over the past billions of years of expansion.[16] In a few years, Alpher and Herman left academia for industry — respectively to the General Electric research laboratories in 1955 and to General Motors in 1956, and at the time no one attempted to observe the relic radiation they predicted.

Although the low-temperature, residual background radiation was eventually discovered, Fred Hoyle had in the meantime developed the Steady State theory that had no relic radiation. Competition between advocates for an everlasting Universe and one with a hot beginning continued for decades.[17] Astronomers were convinced that the Universe was changing over millions and billions of years, but they could not agree on whether or not it exploded from a dense, primeval state, which we now know as the *Big Bang*. Two outspoken proponents of these different perspectives were George Gamow and Fred Hoyle.

The Turbulent Lives of George Gamow and Fred Hoyle

Georgiy Antonovich Gamow began life on March 4, 1904 at the port city of Odessa in the Russian Empire (now in Ukraine), on the northwestern shore of the Black Sea. His parents were high-school teachers; his grandfather on his father's side was a garrison Commander in the Russian Imperial Army, and his mother's father was Archbishop of Odessa.

George's early years were unsettling. His mother died, when he was just 9 years old, and during the subsequent World War I (1914–1918), his retired father had to work as a janitor so they could have something to eat.[18] Some of the family silver was sold to send George to the State University in Petrograd — renamed Leningrad University in 1924 after Lenin's death that year.

Although he rarely attended any university courses, Gamow did listen to lectures on relativity offered by Alexander Friedmann, where he learned about the possibility of expanding or contracting Universes.

In 1928, Gamow received support for a study tour to Göttingen, where he wrote his famed quantum description of the atomic nucleus[19] — when he was only 24 years old. Then after returning to the Soviet Union, George married the attractive physicist Lyubov ("Rho") Vokhminzeva, in August 1931, and the two began several attempts to leave the oppressive country. They first attempted to paddle a kayak across the Black Sea to Turkey, but poor weather ruined the attempt, and they abandoned other possibilities to defect when they realized that they would most likely be caught. After repeated attempts to obtain passports for foreign travel, the couple was eventually permitted to attend the Solvay Conference on Nuclear Physics in Brussels the fall of 1933, and they did not come back.

Marie Curie hosted the couple for two months in Paris, and they moved on to a one-month visit with Ernest Rutherford at the University of Cambridge and a stay with Niels Bohr in Copenhagen for four months. With Bohr's encouragement, Gamow next sought work in America, and found it at the George Washington University in Washington, D.C., where he wrote important papers on the origin of the elements and organized a conference on stellar energy generation that led to solutions of the problem. As he described it, he was just waiting during most of his life, like a spider that sits in the corner of a big web, and when something exciting came he just jumped in, like a spider that quickly goes after something that is caught in his web.[20]

George Gamow has also been remembered for his series of books about Mr. C. G. H. Tompkins and other acclaimed books for the layperson,[21] as well as his unconventional life that included a relentless mockery of science's solemnity and an unrestrained consumption of alcohol.[22]

And this brings us to the other character in this colorful story, Fred Hoyle, who was born June 24, 1915 in his parent's house in Yorkshire, England. His childhood included desperate family poverty, and an unkind teacher who regularly beat her students with a cane. She also boxed their ears with her hand, which Hoyle associated with the eventual loss of hearing in his left ear. While growing up, he acquired a lifelong disrespect for the "all-powerful and all-stupid rampaging monster called 'law'."[23]

In his adolescent years, a different sort of teacher recognized Hoyle's insight, energy and originality, and helped him obtain a scholarship to Emmanuel College at the University of Cambridge, which he entered in 1933 with a pronounced Yorkshire accent, shabby clothes, a jobless father, and a blunt manner. He read mathematics, took the Tripos examinations, and in 1939 married Barbara Clark after just five weeks of knowing each other well enough to have a conversation. The next year Hoyle left Cambridge to help with English radar research, at a time that German aircraft were bombing London.

At war's end in 1945, Hoyle returned to Cambridge, where he spent nearly three decades at the Institute of Astronomy and proposed novel solutions to many significant problems such as the nuclear fusion reactions in stars, the formation of giant stars, the synthesis of elements within stars, and the origin and nature of the Universe.

Fred Hoyle's remarkable creative output included popular astronomy books and lectures, entertaining radio and television broadcasts, and his science fiction classics, *The Black Cloud* and *A for Andromeda*, which helped make Julie Christie a star.

Hoyle became a very public figure as the result of a series of talks on astronomy in the spring of 1949 for the British Broadcasting Corporation, abbreviated BBC. On Saturday evenings millions of listeners tuned into the popular radio broadcasts, which were widely circulated the following year in a short book entitled *The Nature of the Universe*. After ten years of marriage, Hoyle and his wife could now use the lecture payments and book income to buy their first refrigerator.

During his radio broadcast on March 28, 1949, the intelligent and sarcastic Hoyle presented the Steady State theory of the continual creation of matter as the only sensible cosmology, and coined the term *Big Bang* as an expression of derision, without credit to A. S. Eddington who had called it a *Bang* two decades earlier. Eddington thought such an event was unaesthetic, and Hoyle dismissed the unreasonable assumption that matter was created "in one Big Bang at a particular time in the remote past."[24] The Big Bang, he supposed, was an unscientific theory that could never be challenged by direct appeal to observations. The cynical Hoyle also wondered about the scientific credibility of a theory that had been conceived by a Priest and endorsed by a Pope.

Fred Hoyle had a feisty contempt for bureaucracy and orthodox behavior, and regularly spoke out to say just what was on his mind. His combative, pugnacious attitude did not fit in with the traditional, polite behavior of English society, and led to numerous disagreements with his colleagues. They included three decades of public arguments and corrosive relationships with Martin Ryle, the Nobel-prize-winning Cambridge radio astronomer.[25] Their disagreements may have reflected a difference between Hoyle's impoverished family background and Ryle's privileged one — his father was physician to King George VI.

Fred Hoyle did not like religion, and was widely regarded as an atheist. When discussing the beliefs of Christians during one of his BBC radio broadcasts he declared:

"In their anxiety to avoid the notion that death is the complete end of our existence, they suggest what is to me an equally horrible alternative. ... [They] offer an eternity of frustration, [and] have so little to say about how they propose eternity should be spent."[26]

Although the dogmas and miracles in religion were troubling to Hoyle, he retained a religious awe of the Universe and life in it, and had a strong spiritual affinity with a greater power that we do not understand. He sensed that our minds participate in a cosmic intelligence that was somehow involved in the establishment of the Universe and the physical laws that govern it.

As an example, the manufacture of carbon inside stars involves the simultaneous collision of not two helium nuclei but three of them, an exceedingly unlikely situation. Unless the carbon nucleus resonated at a specific energy level, there would not be enough carbon for life to exist. In effect, the resonance increases the cross section for a collision and makes it more likely to occur, like the greater possibility of throwing a ball against a house instead of a tree trunk.

In early 1953, Hoyle visited William Fowler's laboratory at the California Institute of Technology and boldly suggested that there had to be such a resonance of the carbon nucleus, whose energy ought to be at about 7.69 MeV. Ward Whaling, a junior member of Fowler's team, thought there might be something to Hoyle's proposal, and persuaded Fowler to let him look for the effect with equipment no one else was using. When the prediction was verified, Hoyle's name was included on the announcement, but omitted in the published version.[27]

In his autobiographical note for the Nobel Foundation in 1983, Fowler drew attention to Hoyle's important role in our understanding of element-producing nuclear reactions in stars, stating that: "The grand concept of nucleosynthesis was first definitely established by Hoyle in 1946. After Whaling's confirmation of Hoyle's ideas I became a believer."

Although Fowler probably meant his belief in stellar nucleosynthesis, his team's discovery of the excited state of carbon was nevertheless important in turning Hoyle's attention to Design by a Creator God. Hoyle had been involved in the Steady State hypothesis for the expanding Universe, and at the time he was criticized for atheism, but the discovery of his predicted resonance seems to have changed his mind about all that.

When discussing the unlikely origin of carbon, Hoyle commented in 1959 that:

> "If this were a purely scientific question and not one that touched on the religious problem, I do not believe that any scientist who examined the evidence would fail to draw the inference that the laws of nuclear physics have been deliberately designed."[28]

And when writing about the remarkable energy level of the carbon nucleus in subsequent decades, he reaffirmed that: "A 'super-intellect' has monkeyed with physics, as well as with chemistry and biology;"[29] and that:

> "Either our existence is a freakish accident, or the laws of physics were deliberately arranged by some agent to permit our existence."[30]

On New Years Day 1972 Hoyle received a knighthood from Queen Elizabeth II for his service to astronomy, and in the same year his disagreements with the entrenched scientific bureaucracy and political University figures led to his resignation from the Plumian Professorship at the University of Cambridge. In the meantime, he declared that there might be something wrong with his Steady State theory.

After-Glow of the Big Bang

In 1965 Fred Hoyle recanted the Steady State theory and withdrew his objections to the Big Bang hypothesis, even though he never did think it was right.[31] What led Hoyle to renounce his previous outspoken support of the Steady State? The 3-degree relic background radiation of the Big Bang was discovered that year, and it was not predicted by the continual creation theory. The background radiation instead provided strong observational evidence for a dense, primeval origin of the observed Universe.

The discovery of this faint after-glow of the Big Bang was an unexpected outcome of experiments designed for other purposes at the Bell Telephone Laboratories. Their engineers had invented a horn-reflector antenna to detect weak microwave signals from a satellite by rejecting radiation from the ground surrounding the antenna. It was the most sensitive antenna that existed in the world at the time, and it worked just fine in tests carried out in 1962 with instruments aboard *Telstar*, the first working communications satellite.

These tests demonstrated the feasibility of transmitting signals from a place on Earth to an overhead satellite and back. But within one year the Communications Satellite Corporation, abbreviated Comsat, had been created, and the Bell Telephone Laboratories' parent company, the American Telephone and Telegraph Corporation, was legislated out of the international communications-satellite business to avoid a monopoly. So the Bell Telephone Laboratories had this superb horn antenna and detection equipment that were no longer needed, and Arno A. Penzias and Robert W. Wilson decided to use them to make accurate measurements of the intensity of several extragalactic radio sources at a microwave wavelength of 7.34 centimeters or a frequency of 4080 MHz.

Bell Telephone Laboratories hired Arno and Bob with the understanding that they would help with satellite communication, and would be free to work half time on astronomy. They both also had Ph.D. degrees with an astrophysics thesis, and realized there was too much unexpected noise interfering with the detected microwave radiation of known cosmic objects.

They found a persistent, ubiquitous, unvarying and totally unanticipated noise source that contributed an antenna temperature of 3 degrees kelvin, or just three degrees above absolute zero, the coldest possible temperature. The unexpected noise was equally strong in all directions, wherever they pointed the antenna, and independent of the time of the day and of the year. Even a seemingly empty part of space emitted the persistent radiation, and it had no dependence on the location of any known cosmic radio source.

Penzias and Wilson only thought about a cosmological explanation for their result after coming in contact with Princeton Professor Robert H. Dicke, who visited them with his coworkers at the Bell Telephone Laboratories Crawford Hill research facility near Holmdel, New Jersey, located just 40 kilometers (25 miles) from Princeton. The two groups agreed to the publication of two adjacent letters in the *Astrophysical Journal*. The Bell Telephone Laboratories scientists would report on their serendipitous discovery observations in a one-page report, with the modest title *A Measurement of Excess Antenna Temperature at 4080 MHz*,[32] while the Princeton group would write about their possible cosmological interpretations of it in a companion paper, which had a more ambitious title of *Cosmic Blackbody Radiation*.[33] It argued that the ubiquitous and unvarying noise was the first definitive evidence for the Big Bang.

Sometime between the submission and publication of both papers, Walter Sullivan obtained copies and announced the discovery on the front page of the *New York Times*. The headline, on May 21, 1965, read: *Signals Imply a 'Big Bang' Universe*, which told it all. It is this newspaper article that apparently led Robert Wilson to take the cosmology interpretation seriously.

The Nobel Prize in Physics for 1978 was half-awarded to Arno A. Penzias and Robert W. Wilson "for their discovery of cosmic microwave background radiation".

It is now generally accepted that the three-degree radiation is the faint, cooled after-glow of the Big Bang. This remnant radiation is now known as the *three-degree cosmic microwave background radiation*. It is called a *background radiation* because it originated before the stars and galaxies were formed and lies behind them. The galaxies are moving away from the background, like a flock of startled birds in the sky.

We are immersed within the radiation and participating in the explosion. The American poet Robinson Jeffers, whose brother was an astronomer at Lick Observatory, has captured the essence of this Great Explosion with:

> "Nothing can hold them down; there is no way to express that explosion;
> all that exists
> Roars into flame, the tortured fragments rush away from
> each other into all the sky, new Universes
> Jewel the black breast of night; and far off the outer nebulae
> like charging spearmen again
> Invade emptiness."

And perhaps alluding to Lemaître's fireworks Universe, Jeffers' poem includes the lines:

> "No wonder we are so fascinated with fireworks
> And our huge bombs: It is a kind of homesickness perhaps for
> the howling fireblast that we were born from."[34]

Confirming the Discovery

Dicke's research group had been preparing to look for the relic radiation when they were scooped by Penzias and Wilson, so they were able to quickly detect it with their own equipment, and a host of instruments were also sent aloft in high-altitude balloons to get a clear view of the background radiation. Within a decade, the observed spectrum looked very much like that expected from a thermal radiator, or black body, and convinced nearly everyone that it might have cosmological implications. In 1982 Robert Wilson, for example, recalled: "We felt that, at least until they [the theoreticians] had a chance to think about our results, we shouldn't go out on a theoretical limb that we couldn't support. For me, the last nail in the coffin of the Steady-State theory wasn't driven in for quite a while — not until the black body curve was really verified."[35]

Nevertheless, more definitive observations had to be carried out from a satellite with instruments specifically designed to study the cosmic background radiation. They would determine if it has a precise black body spectrum or only an approximation of one, and might detect the temperature fluctuations from place to place that were needed to explain the visible lumpiness of the distribution of galaxies that eventually formed.

Children of the Space Age

On October 4, 1957, the Soviet Union used an intercontinental ballistic missile to launch the first artificial Earth satellite, named *Prosteyshiy Sputnik 1*, the first simple satellite. In shocked response, the United States Congress established the civilian National Aeronautics and Space Administration, abbreviated NASA, to exercise control over activities in space devoted to peaceful purposes for the benefit of all mankind. But threats to world peace also preoccupied Congress, so the very act that established NASA assigned the Department of Defense responsibility for military interests in space.

The Space Age had thus begun, triggered by a Cold War rivalry between the United States and the Soviet Union to gain superiority in rocket, satellite, and space technology.

The competition was won when NASA's *Apollo* flights carried men to the Moon, which demonstrated the superiority of American purpose and technology in space. Over 500 million people around the world watched the televised landing of *Apollo 11* on July 20, 1969, when Neil A. Armstrong took the first human step on the Moon. Although public interest in NASA's space program, and congressional funding of its activities, diminished when the Soviet Union fell apart and American commercial interests and other government agencies gained access to space, NASA's responsibility for space vehicles was nevertheless extended to include rockets, the *Space Shuttle*, space platforms, and space telescopes such as the *Hubble Space Telescope* and the *Spitzer Space Telescope*.

Two young astronomers grew up and matured in this climate, experiencing the Cold War and NASA's triumphs. These children of the Space Age are John C. Mather, born on August 7, 1946, and George Smoot, born on February 20, 1945. They both played key roles in the development, launch and use of a NASA satellite devoted to the study of the cosmic background radiation from above the Earth's obscuring atmosphere. It is named the *COsmic Background Explorer*, abbreviated *COBE*.

Out in space, the instruments aboard *COBE* were not hampered by looking through our planet's atmosphere. The satellite was always surrounded in every direction by the pervasive cosmic microwave background radiation, which emitted far more microwave energy than all the stars and galaxies put together.

Mather's instrument was specifically designed to measure the radiation spectrum, or distribution of radiation intensity as a function of wavelength, to see if it precisely describes the black body spectrum of a perfect thermal radiator. Such a black body absorbs all thermal radiation falling upon it and reflects none — hence the term "black." It radiates energy at a wide range of wavelengths, but with an intensity that is greatest at a wavelength that depends on the temperature. The expansion of the Universe would preserve the black body spectrum of the cosmic background radiation for all time. No process can destroy its shape, but the location of maximum intensity will stretch to longer and longer wavelengths as time goes on and the radiation gets colder.

In the present epoch, with a temperature of about 3 degrees above absolute zero or 3 degrees kelvin, the black body radiation intensity peaks at a wavelength of 0.001 meters, or 1 millimeter. Unfortunately, the Earth's atmosphere absorbs cosmic radiation at this short wavelength.

Mather and John Boslough have provided a fascinating description of the long, trying ordeal involved in doing science from space, particularly with *COBE*.[36] During the previous twenty-five years, each instrument on *COBE* had been preceded by numerous tests of earlier versions lofted in high-altitude balloons or rockets. That involved the dedicated, hard work of large teams of scientists, engineers and managers who faced scientific, technological and bureaucratic obstacles. Even lawyers were involved.

Instruments designed for NASA's specific scientific objectives have to win a peer-reviewed *Announcement of Opportunity*, with stiff competition from other proposals. Once the *COBE* proposal won, the travail had just begun, as the launch vehicles were changing and the instruments needed to be reconfigured. Altogether the *COBE* venture was quite a different kettle of fish from Penzias and Wilson's unexpected discovery, and there was nothing serendipitous about it. True experimenters, the *COBE* team was hunting for something that might or might not be there, and they were careful to confirm their results before announcing them.

On January 13, 1990, less than two months after *COBE* went into orbit but a quarter of a century after the discovery of the background radiation, John C. Mather reported the combined results of the first *COBE* spectral measurements at an American Astronomical Society meeting near Washington, D.C. The spectrum fit the Planck black body curve with error bars of 1%. Later work using the whole data set improved the precision to one part in 20,000 (Fig. 13.1), establishing a temperature of precisely 2.725 degrees kelvin, with an uncertainty of 0.001 degrees kelvin.[37] The presentation caused the audience to break into a standing applause, which you usually see only at the end of a beautiful performance of music.

Such a spectrum could not have happened in the Universe as it is now. Atoms, interstellar material, planets, stars and galaxies all now have a very different temperature than the background radiation. And to put it another way, the observed spectrum is proof that the observable Universe did expand from a very hot, dense state in the past, when matter and radiation were at the same temperature.

It took more than two more years for significant results to be accumulated by the second (of three) *COBE* instruments. George Smoot announced them at the April 23, 1992 meeting of the American Physical Society, but a

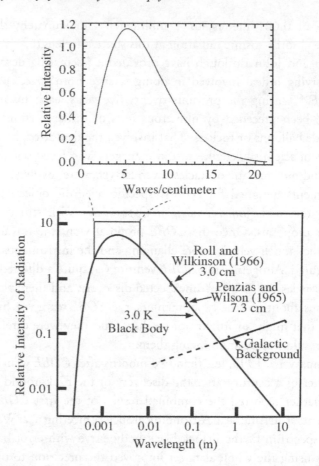

Figure 13.1 Background spectrum The intensity of the cosmic microwave background radiation plotted as a function of wavelength. Pioneering measurements are compared to the expected spectrum of a three-degree black body and radiation from our Galaxy (*bottom*). The full spectrum at millimeter wavelengths (*top*) was obtained from instruments aboard the *COsmic Background Explorer*, abbreviated *COBE*, in late 1989. It corresponds to a black body with a temperature of 2.725 degrees kelvin.

press release had been issued two days earlier from his home institution, the Lawrence Berkeley Laboratory in California. That caused a considerable uproar, since Smoot had broken his signed policy agreement involving prior approval of all papers and press releases by both the full *COBE* science team and NASA officials. His attempts to garner credit for himself appalled everyone involved; but the discomfort was partly alleviated by Smoot's subsequent letter of apology.

After subtracting out the known microwave emission of the Milky Way and using mathematical averaging techniques on about 100 million observations, the *COBE* team found that the temperature varies ever so slightly over large angular sizes.[38] The sensitive instrument detected minute temperature differences no larger than a hundred-thousandth, or 10^{-5}, of a degree kelvin. These fluctuations in the background radiation must have provided the beginning seeds from which today's material Universe of stars and galaxies grew. Otherwise they wouldn't exist.

This was a very important finding. The observed temperature fluctuations portray ripples in the fabric of space-time that existed prior to the formation of the first stars and galaxies, which had to coalesce out of the low-level fluctuations. Mather and Smoot were jointly awarded the 2006 Nobel Prize in Physics "for their discovery of the black body form and anisotropy of the cosmic microwave background radiation."

The *COBE* instrument only probed the largest angular scales, and there was much to be learned by observing smaller scales and by measuring the temperature fluctuations with much greater sensitivity. In the subsequent decade more than 20 experiments from the ground and balloons provided sharper focus for localized regions of the background radiation. Nevertheless another satellite experiment was still needed that would scan the entire sky with enormously improved accuracy, precision, and sensitivity.

At about this time, NASA was faced with intense public scrutiny, having launched the *Hubble Space Telescope* with a faulty mirror, lost the billion dollar *Mars Observer* spacecraft, and blown up at least one *Space Shuttle*. So NASA administrator Daniel S. Goldin decided to cut the losses, and to begin doing things "better, faster and cheaper" using mid-sized rockets to launch inexpensive missions within a year or two of approval.

Charles L. "Chuck" Bennett, at the Goddard Space Flight Center, and David T. Wilkinson, at Princeton, formed a small team of experts to design a spacecraft instrument that could measure angular variations in the temperature of the background radiation within the budget cap of $70 million in 1994 dollars. This cost threshold was raised to $95 million in the late 1990's when two other "cheaper" missions failed, but still a relatively low-cost payload; it did not include the launch costs of about $55 million and the operations costs of about $1.5 million per year.

For a two-decade period beginning in 1993, Bennett led team efforts for the design, proposal, operations, and findings of the key mapping instrument

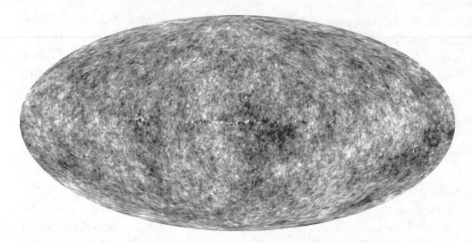

Figure 13.2 Map of the young Universe An all-sky view of the three-degree cosmic micro-wave background radiation emitted from the Universe just 390,000 years after the Big Bang that occurred 13.7 billion years ago. These temperature fluctuations provided the seeds from which galaxies subsequently grew. (Courtesy of the NASA/*COBE* and NASA/*WMAP* Science Teams.)

known as the *Microwave Anisotropy Probe*, or *MAP* for short. It was approved in mid-1996, and launched on June 30, 2001, with a name change in early 2003 to *WMAP*, or *Wilkinson Microwave Anisotropy Probe* to honor Wilkinson after his death.

The most detailed and precise map yet provided for the background radiation was obtained from *WMAP* (Fig. 13.2). The observed temperature fluctuations were used to infer the gravitational pull that caused them, and to reliably tell us what the young Universe was made of, how dense it was, and what forces were at play at the time.[39]

They have shown that the amount of "ordinary" matter, the baryonic kind that makes up atoms, is a small fraction of the total material content of the Universe, and that dark, invisible non-baryonic matter is about five times more abundant than ordinary baryonic matter. Moreover, the combined gravitational pull of both kinds of matter is not enough to stop the expansion of the Universe in the future. The time of the Big Bang explosion that gave rise to the expanding Universe was also dated to 13.7 billion years ago, provided that the Universe has been expanding at a constant rate for all that time.

14. The Origin of the Chemical Elements

"Ay, for 'twere absurd
To think that Nature in the Earth
bred gold
Perfect i' the instant: something
Went before
There must be remote matter."

Benjamin "Ben" Johnson (1612)[1]

Where did the Elements Come from?

In the early 19th century the English chemist John Dalton proposed an atomic theory of matter in which chemical compounds are formed from the combination of atoms of definite and characteristic weight.[2] Dalton knew that hydrogen was the lightest element, so he gave it an atomic weight of one, and he assigned larger numbers to the other atoms. These elements are now arranged in the periodic table by their atomic number and chemical properties.

Most of the natural chemical elements found on the Earth are exceptionally durable and long-lived. When the substance they occupy wears away, its elements disassemble and become redistributed in something else. Nothing lasts, everything changes, but the constituent elements mostly just get moved around.

So we naturally wonder how the abundant, long-lived, stable chemical elements, like hydrogen, carbon, nitrogen and oxygen, were made? As suggested by their durability, it takes an unusual amount of energy to produce or transform these chemical elements, and that is why ancient alchemists never succeeded in transforming lead into gold within their laboratories. Very high temperatures are required to energize the nuclear reactions that create chemical elements, and they naturally occurred only in the first moments of the expanding Universe or subsequently in the hot interiors of stars.

Clues to the nuclear reactions that gave rise to the elements in these two locations were provided by the discovery that the same chemical elements are found throughout the Cosmos, and by observations of how the relative abundances of these elements depend on their weight.

Universal Chemical Elements

Dark absorption lines or bright emission lines reveal the presence of an element in a cosmic object. Spectroscopic observations of these line features identify the element and only that element. The spacing and relative strengths of the hydrogen lines, first described by Johann Balmer in 1885, is an example.[3]

About a quarter of a century before that, Gustav Kirchhoff, at the University of Heidelberg, and his chemist colleague Robert Bunsen, were identifying the chemical composition of different substances by the spectral lines they emitted when burned. Since most of the Sun's spectral lines could be identified with lines known from their laboratory experiments, Kirchhoff and Bunsen concluded in 1860 that the Sun and Earth contain the same chemical elements, with the same relative abundance.[4] For a time, astronomers even talked of an iron Sun, owing to the great abundance of iron in the Earth and in the solar line spectra.

Stars of different spectral types showed conspicuous lines of various elements, however, suggesting that some stars might have different chemical compositions. But it was eventually realized that the presence or absence of specific spectral lines depends on the physical conditions in a star's outer atmosphere and does not necessarily indicate its chemical composition.[5]

As far as astronomers could tell, the same chemical elements are present in the Earth and the stars, but in different proportions, or relative amounts. The most dramatic differences are in the substantial relative amounts of hydrogen and helium in stars.

Cecilia H. Payne and Stars Dominated by Hydrogen

The English-American astronomer Cecilia H. Payne was born in 1900 at the foot of the Buckingham Chilterns in England, where her ancestors had tilled the soil for at least ten centuries. As a child, she loved the surrounding natural world. Trees were her earliest companions, and the captivating sight of spiders, mimosa, and orchids were still remembered in her later years. Her early education at St. Paul's Girls' School also instilled a love of music, when she came under the spell of her shy, charming teacher Gustav Holst and his newly composed *The Planets*.

Cecilia completed her undergraduate studies on a scholarship to Newnham College at the University of Cambridge as a student in natural science. When recalling her college years, she was fond of quoting the English poet William Wordsworth's lines: "Nature never did betray the heart that loved her," as well as the English biologist T. H. Huxley's advice that the student of Nature should: "Sit down before fact as a little child, be prepared to give up every preconceived notion, follow humbly wherever and to whatever abysses Nature leads, or you shall learn nothing."[6]

Upon hearing a lecture by A. S. Eddington near the end of 1919, about the observational confirmation of the curvature of space-time by massive objects, Payne switched her studies from the life sciences to the physical ones, with the zealous enthusiasm of a religious convert. Later, during a public night at the Cambridge Observatory, she met Eddington and blurted out: "I should like to become an astronomer," which he encouraged.

After completing her undergraduate degree in 1923, Cecilia attended another memorable lecture in London, by the young Harlow Shapley on his discovery of the remote center of the Milky Way. It resulted in her application for graduate study at the Harvard College Observatory, which Shapley directed. When Shapley provided Miss Payne with a fellowship meant to encourage women, she accepted the opportunity and left her family and native country for the Observatory, which became her home for more than 50 years.

The Harvard astronomers were compiling an enormous collection of stellar spectra that were at Cecilia's disposal. Her brilliant doctoral thesis and book on *Stellar Atmospheres*, published in 1925, summarized the existing spectroscopic data for a wide variety of stars, and described how the appearance of their spectral lines is influenced by physical conditions in stellar atmospheres. She showed that the observed line strengths in various stars are mainly due to different temperatures, rather than a difference in their composition, and that the relative abundances of the chemical elements are similar in virtually every bright star. Her results also showed that the Sun and many other stars contain elements commonly found on the Earth, such as carbon, silicon and iron, and in roughly the same proportions.

Cecilia Payne was the first person to earn a doctorate degree in astronomy from Radcliffe College, which is now part of Harvard University. For many subsequent years, she remained upset about the complete lack of recognition from either Harvard or Radcliffe, who both failed to adequately acknowledge her important spectroscopic investigations. But she eventually became the first female to be promoted to Full Professor from within the Faculty of Arts and Sciences at Harvard University, and was later promoted to Chairperson of the Department of Astronomy.

Cecilia benefited from collaborations with other women at the Harvard College Observatory, such as Annie Jump Cannon, Henrietta Swan Leavitt and Antonia C. Maury. They were all unmarried, received no faculty appointments, and suffered from some sort of physical handicap. Annie Cannon and Henrietta Leavitt were both deaf, and Antonia Maury was extremely homely.

In 1931 Cecilia Payne became an American citizen, and in 1934 she married the Russian-born astronomer Sergei Gaposchkin. They settled in Lexington, Massachusetts, where they raised three children. The family was active in the First Unitarian Church in Lexington, where Cecilia taught Sunday School class and advocated the disappearing art of *Biblical* quotation.

In retrospect, Cecilia's most significant finding was one she could not herself believe. Although the observed stellar disks contain elements that are commonly found on the Earth, the lightest elements, hydrogen and helium, are vastly more abundant in stars than they are on our planet. As Cecilia described the difference:

"The outstanding discrepancies between the astrophysical and terrestrial abundances are displayed for hydrogen and helium. The enormous abundances derived for these elements in the stellar atmosphere is almost certainly not real. Probably the result may be considered, for hydrogen, as another aspect of its abnormal behavior... The lines of both atoms appear to be far more persistent, at high and low temperatures, than those of any other element."[7]

Her analysis indicated that hydrogen was as much as a million times more abundant in the Sun's outer atmosphere than it is on the Earth, and she just could not understand how that might happen. At the time, all publications at the Harvard College Observatory had to be approved by its director Harlow Shapley, and he most likely encouraged her to play down the glaring hydrogen discrepancy.

In just four years, the influential Princeton astronomer Henry Norris Russell had nevertheless confirmed and extended Cecilia Payne's conclusion that hydrogen is the overwhelmingly predominant element in the Sun's outer atmosphere.[8] And within a few more years, the Danish astronomer Bengt Strömgren, an expert on stellar structure, showed that the physical conditions within stars requires that the great hydrogen content exists throughout their interiors, not just in their outer atmospheres.[9] The enormous abundance of hydrogen then played a central role in discovering how the Sun shines, by the nuclear conversion of hydrogen into helium in its core [See Chapter 9].

Relative Abundances and Synthesis of the Elements

An important clue to understanding where the elements came from is obtained from their relative abundances, initially studied by chemists rather than astronomers. The American chemist William D. Harkins, for example, found an important clue when he noticed that elements of low atomic weight are more abundant in meteorites than those of high atomic weight. These features led Harkins to conjecture in 1917 that the relative abundances of the elements depend on nuclear rather than chemical properties and that heavy elements must have been synthesized from the lighter ones, starting from hydrogen.[10]

And A. S. Eddington wrote in 1920 that: "I think that the suspicion has been generally entertained that the stars are the crucibles in which the lighter atoms which abound in the nebulae are compounded into more complex elements."[11]

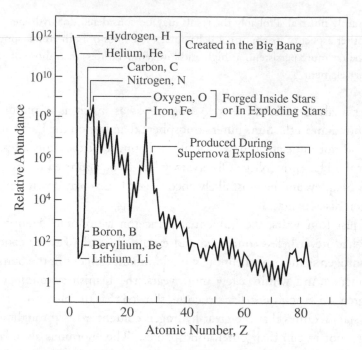

Figure 14.1 **Abundance and origin of the Sun's elements** The relative abundance of the elements in the solar photosphere, plotted as a function of their atomic number, Z. Hydrogen, the lightest and most abundant element in the Sun, and most of the helium, were formed about 14 billion years ago in the immediate aftermath of the Big Bang that led to the expanding Universe. All the elements heavier than helium were synthesized in the interiors of stars that no longer shine, and then wafted or blasted into interstellar space where the Sun subsequently originated. (Data courtesy of Nicolas Grevesse.)

But it wasn't until 1956 that Hans E. Suess and Harold Clayton Urey, at the University of Chicago, published detailed abundance data for meteorites that suggested that most elements were made inside stars (Fig. 14.1).[12] The systematic decline in abundance of elements with increasing atomic weight was explained by the fact that the nuclear reactions in the most common stars produced only a small amount of heavy elements.

These element-building processes are now termed *nucleosynthesis*, which describes the synthesis of atomic elements from their sub-atomic, nuclear components, the neutrons and protons rather than whole atoms. Some of the elements were made within stars, and thrust back into space. This is called *stellar nucleosynthesis*, and it accounts for the production of heavy elements.

But where did the abundant lightest elements hydrogen and helium come from? They were made in the earliest stages of the expanding Universe, before the stars were formed. It is known as *big-bang nucleosynthesis*.

Creating Elements in the Big Bang

Support for big-bang nucleosynthesis arose as a byproduct of the development of the atomic bomb during World War II (1939–1945). George Gamow knew how elements were being produced by chain reactions during the detonation of these bombs, and shortly after the war, in 1946, he concluded that all of the elements originated by similar explosive nuclear reactions during the Big Bang, the most energetic and hottest explosion of them all.[13]

Working with Ralph A. Alpher, his graduate student at George Washington University, Gamow proposed that the original substance of the Universe was a hot, highly compressed nuclear gas, which they named *ylem*, the material from which the elements formed. This ylem was supposed to consist only of neutrons at a very high temperature of about 10 billion degrees kelvin.

The cosmic ylem cooled as the Universe expanded, and some of the neutrons would have decayed into protons. According to Gamow and Alpher successive captures of neutrons by protons led to the formation of the elements when the temperature dropped. The probability of a neutron being captured had been measured in the Los Alamos Atomic Energy (Manhattan) Project, and using these data, which were declassified after World War II, they obtained a reasonable fit to the observed decline of element abundance with increasing weight.

Their novel idea was published in 1948, in a paper entitled "The Origin of the Chemical Elements," with Hans Bethe added as an author by the fun-loving Gamow, even though Bethe contributed nothing to the research or the writing. This was a pun on the first letters of the Greek alphabet — alpha, beta and gamma or α, β, and γ — for Alpher, Bethe and Gamow.[14] Bethe, who was one of the reviewers of the article when it was submitted, enjoyed the joke. To add to the fun, the paper was published on April Fool's Day.

For several years, Alpher teamed up with Robert Herman, a fellow employee at the Johns Hopkins University, to continue these calculations. With immodest humor, they used "divine creation curves" to specify the decrease in temperature and density as the expanding Universe cooled and dispersed.

Modern computations have conclusively demonstrated that all of the hydrogen and deuterium, and most of the helium, that are found in the Cosmos today were synthesized in the immediate aftermath of the Big Bang.[15] These calculations also indicated how much matter is now in the Universe in both visible and invisible form.

So Gamow and his colleagues were partly right. The lightest and most abundant element, hydrogen, and therefore the majority of atoms that are around today, was indeed formed at the dawn of time before the stars even existed, in the immediate aftermath of the Big Bang that produced the expanding Universe. All of the hydrogen that is now found in stars and interstellar space was created back then, about 13.7 billion years ago, and so was most of the helium, the second-most abundant element. Although helium is synthesized within stars, the amounts of helium now observed in the Cosmos are much greater than the amounts that could have been produced in all the stars in the Universe over its entire lifetime. Most of it had to be created in the early stages of the Big Bang. So every time you buy a floating balloon, which has been inflated with helium, you are getting atoms made many billions of years ago.

To put it another way, the close fit between the calculated and measured cosmic abundances of the light elements provides strong evidence that the Big Bang occurred and that the observed Universe had such a beginning.

But Gamow was also partly wrong, for all of the chemical elements were not created in the Big Bang, only the abundant lighter ones were. The creation of carbon, the next most abundant atom after helium, requires the simultaneous collision of three helium nuclei, known as a *three-body reaction*. By the time that the expanding Universe became sufficiently cool for atomic elements to form, at about 1 billion degrees kelvin, the material was so dispersed, thinned out, and cooled down that only helium could be produced in abundance.

The cores of stars are the only places that are dense enough and hot enough to synthesize elements heavier than helium, over long intervals of time.

Creating Elements in Stars

Hydrogen is converted into helium within the cores of stars, and these nuclear reactions provide the energy that supports these stars and makes them shine.[16] These reactions are known as *hydrogen burning*. The important remaining question was where the heavier abundant elements, such as carbon, nitrogen and oxygen, came from. They were produced in the late stages of the lives of exceptionally massive and luminous stars.

Once the hydrogen in a star's central regions has been converted into helium, the core contracts and heats up, releasing energy that causes the outer envelope of the star to expand into the vast giant structure. The helium ash produced by hydrogen burning serves as nuclear fuel in the hot, compact cores of giant stars.

The Estonian astronomer Ernst Öpik and the American astronomer Edwin E. Salpeter nearly simultaneously showed how carbon nuclei could be created in this situation during triple collisions of helium nuclei.[17] The release of energy by fusing helium into carbon within a star is known as *helium burning*.

The English astrophysicist and cosmologist Fred Hoyle was one of the first to propose that many other chemical elements were forged under several different situations inside stars, and he collaborated with the American nuclear physicist William Fowler in investigating these possibilities.[18] The duo also found time to climb the Munros in the Scottish Highlands, which kept them fit and according to Fowler renewed his soul.

William A. Fowler and Stellar Nucleosynthesis

William "Willy" Fowler was raised in the railroad center of Lima, Ohio, where he acquired a lifelong fascination with steam locomotives. His paternal grandfather was a coal miner who came to Pittsburgh from Scotland, his maternal grandfather was a grocer who immigrated to Pittsburgh from Ireland, his father was an accountant, and his family was not affluent.

On graduation from high school, Willy studied Engineering Physics at Ohio State University, where he waited on tables, washed dishes, and stoked furnaces for meals. He was admitted to graduate school at the California Institute of Technology in 1933 with an assistantship of room, board and tuition, and was assigned to work at the Kellogg Radiation Laboratory where he and other graduate students made experimental measurements of nuclear reactions at very low energies. [The laboratory was named for the American "corn flakes king" Will Keith Kellogg who funded it.]

These investigations were stopped during the Second World War (1939–1945), when the people at Kellogg developed, tested and produced small rockets for the Navy and constructed non-nuclear components of the first atomic bombs. Heavy elements were created during the tests of these bombs, which indicated that these elements did not have to be produced at the Big-Bang beginning of the observable Universe.

After the war, Fowler and his colleagues at Kellogg stayed away from the competitive and lucrative construction of high-energy accelerators used to study elementary particles, and instead resumed laboratory measurements of nuclear processes involved in stars. These experiments and their interpretation continued at the California Institute of Technology for half a century, with a minimum teaching load, little academic bureaucracy, and the support of the Office of Naval Research and the National Science Foundation.

Fowler's team carried out bombardments of nuclei at relatively low energies, which excited the nuclei but did not tear them into pieces. The group also converted the laboratory nuclear interaction cross sections into reaction rates inside stars, which no theorist could predict.[19] In this way, they helped determine how energy is generated within stars by converting light elements into heavier ones, and also accounted for the cosmic creation of the elements. These discoveries were considered so important that William Fowler was awarded the 1983 Nobel Prize in Physics for them.

We Belong to the Stars

Fowler and Fred Hoyle teamed up with Margaret and Geoffrey Burbidge to write an overview of the synthesis of elements in stars. It emphasized the details of the cosmic abundance of the elements and explained both the abundant elements and those that are relatively under-abundant or just not there. This influential review became widely known by the acronym B^2FH.[20]

The B²FH paper showed how all of the elements from carbon, of atomic number six, to uranium of number ninety-two, can be created by nuclear processes in stars starting with hydrogen and helium, which were produced in the Big Bang. The heavier elements were forged during slow, successive nuclear burning stages in stars and the abrupt explosions of stars near the ends of their lives.

So stars are the crucibles in which all the chemical elements heavier than hydrogen and helium were formed. These elements have been created by nuclear reactions in stars during their ongoing lives and then blown or blasted into space by dying stars, gathered by gravity, and collected into the next generation of stars and their planets.

We are here today because former stars forged the chemical elements necessary for life within their nuclear furnaces. You are composed of these elements, assemblages of former star stuff. So Walt Whitman was right to say: "I believe a leaf of grass is no less than the journey work of stars,"[21] and Joni Mitchell was right on when she sang:

> "We are stardust
> We are golden."[22]

There is one remaining mystery that no astronomer has solved. We just do not know exactly how the observable Universe came into being in the first place or where it came from. A precise knowledge of this very beginning is hidden behind a closed door.

References and Notes

Many of these articles have been reproduced with commentaries about them by the author and Owen Gingerich, and subsequently collected and extended by Marcia Bartusiak over a broader range of time and with her own introductions. These sources are designated in the references by the acronyms:

LGSB = Kenneth R. Lang (1941–) and Owen Gingerich (1930–): *A Source Book in Astronomy and Astrophysics 1900–1975*. Cambridge, Massachusetts: Harvard University Press 1979.

MBARK = Marcia Bartusiak (1950–): *Archives of the Universe: 100 Discoveries that Transformed Our Understanding of the Cosmos*. New York: Vintage Books, Random House 2004.

Part I. Everything Moves

1. The Earth Moves

1. John Milton (1608–1674): *Paradise Lost*. London: Samuel Simmons 1667, *Book VIII, The Argument*, lines 79 to 84. New York: Macmillan Pub. Co. 1993 [ed. Roy C. Flannagan (1897–1952)], p. 433.
2. The ancient Roman philosopher Marcus Cicero declared more than two thousand years ago that: "When we lift our eyes to the heavens and consider the celestial bodies, what can be more manifest, more obvious, than there must be a supreme intelligence by which they are ordered ... a deity who is ever present and all-powerful. Marcus Tullius Cicero (106 BC–43 BC): *De Natura Deorum (On the Nature of the Gods), Book II, Number 2*, 45 BC. English translation by Hubert M. Poteat, Chicago: University of Chicago Press 1950, p. 227.
3. Robert Millikan (1868–1953): *Evolution in Science and Religion, The Terry Lectures for 1927*. New Haven, Connecticut: Yale University Press 1927, p. 79.
4. Around 1510, Copernicus was privately circulating a manuscript in which the Earth and other planets were placed in uniform circular motion about a central

Sun. It was entitled *Nicolai Copernici de hypothesibus motuum coelestium a se constitutes commentariolus*, or *A Little Commentary About Nicolaus Copernicus' Hypotheses of Celestial Motions*, and is commonly referred to as the *Commentariolus*, or *Little Commentary*. Though never intended for printed distribution, hand-written manuscript copies were distributed to Copernicus' close friends. The surviving copies of the *Little Commentary* indicate that it contained all his fundamental new ideas about the nature of heavenly motions, which also appeared in his longer, more influential book *De revolutionibus orbium coelestium* or *On the Revolutions of the Celestial Bodies*, commonly referred to as the *De revolutionibus*. It was published 33 years later, in 1543, just in time for the dying author to hold the pages that would make him immortal. See MBARK, pp. 51–60.

5. The author's synthesis of the English translations given by Arthur Koestler (1905–1983): *The Sleepwalkers: A History of Man's Changing Vision of the Universe*. New York: Penguin Books 1964, pp. 205, 206, and James MacLachlan (1928–): *Nicolaus Copernicus: Making the Earth a Planet*. Oxford: Oxford University Press 2005, p. 63.

6. Owen Gingerich (1930–): "The Astronomy and Cosmology of Copernicus," in his *The Eye of Heaven: Ptolemy, Copernicus, Kepler*, New York: American Institute of Physics 1993, pp. 181–182. His translation of Nicolaus Copernicus (1473–1543): *De revolutionibus orbium coelestium* (Nuremberg, 1543). Book 1, Chapter 10. Also in Owen Gingerich (1930–): *God's Universe*. Cambridge, Massachusetts: Harvard University Press 2006, pp. 17–18 and pp. 86–87.

7. The tiny, wobbling annular parallax shift of a nearby star's position was not definitely observed until 1838, by the German astronomer Friedrich Bessel (1784–1846) for the star 61 Cygni, published in *Astronomische Nachrichten* **16**, 65–96 (1839). See MBARK, pp. 153–159.

8. In 1852 the French physicist Léon Foucault (1819–1868) showed that the plane of oscillation of his swinging pendulum in the Pantheon of Paris slowly shifted in direction as the Earth turned below it. The concept of a rotating Earth has a long history. The Greek astronomer Heraclides Ponticus proposed in the 4th century BC that both the Sun and stars might not be moving, and that the Earth could instead be turning under them once every day. A rotating Earth was also considered and dismissed by both Aristotle and Ptolemy. The fact that the Earth remained motionless was just obvious to everyone. The ground certainly seems to be at rest beneath our feet, providing the *terra firma* on which we carry out our lives. And if the Earth was spinning about every 24 hours, its atmosphere and everything on the planet might be flung off it into surrounding space.

9. Stillman Drake (1910–1993): *Discoveries and Opinions of Galileo*. New York: Anchor Books, 1957, p. 5.

10. The term *spyglass*, or "glass used to spy," was replaced with the Latin *telescopium*, or the English equivalent *telescope*, when Galileo attended a banquet of Rome's Academia dei Lincei in 1611; the word is coined from the Greek roots *tele* for "far off" and *scope* meaning "to look."

11. Galileo Galilei (1564–1642): *Letter to the Doge of Venice, 1609*, located in *Le Opere di Galileo Galilei: Edizione Nazionale* **10**, 250–251. This Letter to the Chief Magistrate of Venice describes the benefits of the telescope in war, including its ability "to discover at a much greater distance than usual the hulls and sails of the enemy, so that for two hours and more we can detect him before he detects us and, distinguishing the number and kind of his vessels, judge his force, in order to prepare for chase, combat or flight." [English translation by Albert Van Helden (1940–), in his introduction to Galileo's *Sidereus Nuncius*, or *The Sidereal Messenger*. Chicago: University of Chicago Press 1989, pp. 7 and 8.]

12. Although few of Galileo's many telescopes survive, he probably used one with an objective lens of about 0.05 meters (2 inches) in diameter to make his startling discoveries. The collecting area of such a lens is roughly fifty times that of the pupil of the unaided eye, which is about 0.007 meters across. The increase in light-collecting power of even this small telescope enabled Galileo to discover a host of previously unknown features in the Universe. As a result, the Cosmos became much richer and more complex than had even been imagined at the time.

13. Owen Gingerich (1930–) and Albert Van Helden (1940–): "How Galileo Constructed the Moons of Jupiter," *Journal for the History of Astronomy* **42**, 259–264 (2011).

14. Galileo Galilei (1564–1642): *Sidereus Nuncius*, Venetiis: Apud Thomam Baglionum 1610. English translation by Albert Van Helden (1940–): *The Sidereal Messenger*. Chicago: University of Chicago Press 1989. Also see MBARK, pp. 76–89. In Prague, the Tuscan ambassador had a copy of *Sidereus Nuncius* delivered to the astronomer Johannes Kepler, imperial mathematician to the emperor Rudolf II. Late that summer, Kepler borrowed a telescope, and used it to verify the existence of Jupiter's moons. He then confirmed Galileo's observations in a short pamphlet in which he also coined the term *satellite* that is now used to designate any object that orbits a planet. See Johannes Kepler (1571–1630): *Narratio de Observatis a se Quatuor Jovis Satellitibus*, or *Narration about Four Satellites of Jupiter Observed* (1610). The four large moons that Galileo discovered are now known as the "Galilean satellites" in his honor; these incredible new worlds have been revealed in fascinating detail from NASA's *Galileo Mission*.

15. John Donne (1572–1631): *An Anatomy of the World: The First Anniversary,* lines 279, 280. London: Samuel Macham 1611.

16. Galileo Galilei (1564–1642): *Istoria e dimostrazioni intorno alle macchie solari e loro accidente,* or *History and Demonstrations Concerning Sunspots and Their Phenomena* (1613) and commonly called *Letters on Sunspots,* reproduced in English translation by Stillman Drake (1910–1993): *Discoveries and Opinions of Galileo.* New York: Anchor Books, 1957, the quotation is from the first *Letter on Sunspots,* see p. 94 of the Drake translation.

17. Galileo Galilei (1564–1642): *Istoria e dimostrazioni intorno alle macchie solari e loro accidente,* or *History and Demonstrations Concerning Sunspots and Their Phenomena* (1613) and commonly called *Letters on Sunspots.* Reproduced in English translation by Stillman Drake (1910–1993): *Discoveries and Opinions of Galileo.* New York: Anchor Books, 1957, the quotation is from p. 119.

18. Galileo Galilei (1564–1642): Introduction to *Sidereus Nuncius* (1610), English translation by Albert Van Helden (1940–): *The Sidereal Messenger,* Chicago: University of Chicago Press 1989, pp. 17–18.

19. Galileo's third *Letter on Sunspot* (1613), translated by Stillman Drake (1910–1993): *Discoveries and Opinions of Galileo.* New York: Anchor Books, 1957, p. 128.

20. Galileo Galilei (1564–1642): *Letter to Benedetto Castelli,* December 21, 1613. Translated by Stillman Drake (1910–1993): *Galileo at Work: His Scientific Biography.* Chicago: University of Chicago Press 1978, pp. 224–225.

21. Galileo Galilei (1564–1642): *Letter to Madame Christina of Lorraine, Grand Duchess of Tuscany: Concerning the Use of Biblical Quotations in Matters of Science* (written and circulated in 1615, published in 1636), English translation by Stillman Drake (1910–1993): *Discoveries and Opinions of Galileo.* New York: Anchor Books, 1957, pp. 176–216. A marginal note by Galileo attributes the quoted epigram to Cardinal Cesare Baronio, p. 186.

22. Stillman Drake (1910–1993): *Discoveries and Opinions of Galileo.* New York: Anchor Books, 1957, p. 164.

23. This passage is from Galileo's *The Assayer* (1623). English translation in Stillman Drake (1910–1993): *Discoveries and Opinions of Galileo.* New York: Anchor Books, 1957, p. 239. A contemporary saying asks: "How can you fly like an eagle when you work with a bunch of turkeys."

24. Galileo's confession before seven Cardinal-Inquisitors on June 22, 1633, English translation in James MacLachlan (1928–): *Galileo Galilei: First Physicist.* Oxford: Oxford University Press 1997, p. 76.

25. Johannes Kepler (1571–1630): *Astronomia Nova …* (*New Astronomy Based upon Causes, or Celestial Physics, Treated by Means of Commentaries on the Motions of*

the Star Mars, from Observations of Tycho Brahe). Prague: Romanorvm Imperatoris 1609. See MBARK, pp. 67–75.

26. Johannes Kepler (1571–1630): *Kepler: Gesammelte Werke 16*, 37, Max Casper and Walther von Dyck (Eds.): Munich: C. H. Beck 1938, translated by Arthur Koestler (1905–1983): *The Sleepwalkers: A History of Man's Changing Vision of the Universe*. London: Hutchinson 1959, pp. 394, 395.

27. Johannes Kepler (1571–1630): Quotation in James R. Voelkel (1962–): *Johannes Kepler and the New Astronomy*, New York: Oxford University Press 1999, p. 66.

28. Johannes Kepler (1571–1630): *Kepler: Gesammelte Werke 3, dedication*, translated by Arthur Koestler (1905–1983): *Ibid.* p. 325.

29. Johannes Kepler (1571–1630), *Letter to Herwart von Hohenberg*, the Bavarian Chancellor and Kepler's patron, dated 10 February 1605. Reproduced in *Kepler: Gesammelte Werke,* **15**, 145, translated by Arthur Koestler: *The Sleepwalkers: A History of Man's Changing Vision of the Universe*. New York: Penguin Books 1964, p. 345.

2. Gravity Guides Movement and Bends Space-Time

1. Alexander Pope (1688–1744): *An Essay on Man, Epistle II*, lines 19 to 22. London: Printed for J. Wilford 1734.

2. Isaac Newton was born on Christmas Day 1642 according to the Julian calendar then in force in England; the rest of Europe was already using the modern Gregorian calendar, for which the date was January 4, 1643.

3. *The Holy Bible, King James Version, The Revelation of St. John the Divine, Revelation 14*: 4–5.

4. Isaac Newton (1642–1727): "An Account of a New Catadioptrical Telescope, invented by Mr. Isaac Newton," *Philosophical Transactions of the Royal Society*, No. 81 (25 March 1672), pp. 4004–4007.

5. Isaac Newton (1643–1727): "A Letter of Mr. Isaac Newton containing his New Theory about Light and Colours," *Philosophical Transactions of the Royal Society*, No. 80 (19 Feb. 1671/72), pp. 3075–3087.

6. Isaac Newton (1642–1727): King's College Library, Cambridge: *Keynes Manuscript 130 (7)*. Quoted by Michael White: *Isaac Newton: The Last Sorcerer*. Reading, Massachusetts: Addison-Wesley 1997, p. 43.

7. Newton was withdrawn, cautious, and suspicious; disliked exposing his thoughts, beliefs, or discoveries to possible criticism or distasteful priority disputes; and published almost nothing except under extreme pressure from his friends. Most of his extensive and unpublished religious writings have therefore remained unpublished and gone relatively unnoticed until recent times when

they have become available. They have now been released and can be examined online at *The Newton Project*.

8. Herbert Westren (H. W.) Turnbull (1885–1961): *The Correspondence of Isaac Newton, Volume III 1688–1694*. Published for the Royal Society by Cambridge University Press 1961, No. 398, pp. 233–236.

9. Isaac Newton (1643–1727): *Principia, Scholium Generale*, end of *Book III. The System of the World*. This *General Scholium* first appeared in the second edition of the *Principia*, in 1713, and not the first edition published in 1687. English translation by Andrew Motte in 1729, revised by Florian Cajori, Berkeley: University of California Press 1934, pp. 544–546.

10. John Maynard Keynes (1883–1946): "Newton, The Man". In *The Royal Society Newton Tercentenary Celebrations 15 July 1946*, Cambridge, Printed for the Royal Society (London) by the University Press 1947, pp. 27–34.

11. I. Bernard Cohen (1914–2003): "Newton," *Dictionary of Scientific Biography, Volume 10*, New York: Scribner's 1974, p. 64. Also see John Maynard Keynes (1883–1946): *Ibid.*, p. 28.

12. Isaac Newton (1643–1727): *The Principia*, Second Edition of 1713. English translation by I. Bernard Cohen (1914–2003) and Anne Whitman (1937–1984), Berkeley: University of California Press 1999, p. 943. See MBARK, pp. 90–98.

13. William Shakespeare (1564–1616): *Julius Caesar* (1599), *Act II, Scene II*, lines 30, 31. Also called *The Tragedy of Julius Caesar* and *The Life and Death of Julius Caesar*.

14. Edmond Halley (1656–1742): "Astronomiae Cometicae Synopsis," *Philosophical Transactions of the Royal Society of London* **24**, 1882–1899 (1705), *A Synopsis of the Astronomy of Comets*. London: John Senex 1705. Reproduced in MBARK, pp. 99–106.

15. Jan H. Oort (1900–1992): "The Structure of the Cloud of Comets Surrounding the Solar System, and a Hypothesis Concerning its Origin," *Bulletin of the Astronomical Institutes of the Netherlands* **11**, 91–110 (1950). Reproduced in LGSB, pp. 132–137, and MBARK, pp. 440–445.

16. William Herschel (1738–1822): "Account of a Comet," *Philosophical Transactions of the Royal Society* **71**, 492–501 (1781). Reproduced in MBARK, pp. 128–131.

17. Urbain Jean Joseph Le Verrier (1811–1877): "Comparaison des observations de la nouvelle planète avec la théorie déduite des perturbations d'Uranus," *Comptes rendus* **23**, 771 (1846). John Couch Adams (1819–1892): "An explanation of the observed irregularities in the motion of Uranus, on the hypothesis of disturbances caused by a more distant planet; with a determination of the mass, orbit, and position of the disturbing body, "*Appendix to the Nautical Almanac for the Year 1851* (London, 1846), *Monthly Notices of the Royal Astronomical Society* **7**, 149–152 (1847). See MBARK, pp. 160–167.

18. Johann Gottfried Galle (1812–1910): *Letter to Le Verrier*, 1846, manuscript letter in the library of the Paris Observatory, and quoted by Morton Grosser (1942–): *The Discovery of Neptune*. Cambridge, Massachusetts: Harvard University Press 1962.

19. Isaac Newton (1642–1727): In Sir David Brewster (1781–1868), *Memoirs of the Life, Writings, and Discoveries of Sir Isaac Newton, Volume II, Chapter 27*. Edinburgh: T. Constable and Co. 1855.

20. Pablo Neruda (1904–1973): *Those Lives*, lines 4, 5. English translation by Alastair Reid (1926–2014) in his *Isla Negra: A Notebook*, New York: Farrar, Straus and Giroux 1981, p. 133, reproduced by permission of the translator.

21. Albert Einstein (1879–1955): *Autobiographical Notes*, in *Albert Einstein: Philosopher-Scientist* (P. Schilpp, Ed.), New York: Tudor 1949, p. 8.

22. Albert Einstein (1879–1955): From an interview, *Saturday Evening Post*, October 26, 1929. Reproduced by Alice Calaprice (1941–): *The Expanded Quotable Einstein*. Princeton, New Jersey: Princeton University Press 2000, p. 155.

23. Abraham Pais (1918–2000): *Subtle is the Lord: The Science and the Life of Albert Einstein*. Oxford: Oxford University Press 1982, p. 311.

24. Albert Einstein (1879–1955): *Letter to President Franklin D. Roosevelt* on August 2, 1939. Reproduced by Alice Calaprice (1941–) *Ibid*, pp. 374–377. As it turned out, Einstein's concern was real, for the Germans were investigating the possibility of constructing nuclear weapons during World War II (1939–1945), but without success. After the war, Einstein renounced the use of nuclear weapons and urged their international abolishment, but it was too late for that.

25. Albert Einstein (1879–1955): *Letter to V. Besso*, March 21, 1955. Reproduced in *Albert Einstein — Michele Besso Correspondence 1903–1955* [Paul Arthur Speziali (1897–1993), Ed.]. Paris: Hermann 1972, p. 537 and Abraham Pais (1918–2000): *Subtle is the Lord: The Science and the Life of Albert Einstein*. Oxford: Oxford University Press 1982, p. 302.

26. Albert Einstein (1879–1955): *Autobiographical Notes*, in *Albert Einstein: Philosopher-Scientist* [P. Schilpp (1897–1993), Ed.], New York: Tudor 1949, p. 95. Also reproduced by Alice Calaprice (1941–): *Ibid*. p. 220.

27. Albert Einstein (1879–1955): "What I Believe," *Forum and Century* **84**, 193–194 (1930), quoted by Max Jammer (1915–2010): *Einstein and Religion*. Princeton, New Jersey: Princeton University Press 1999, p. 73.

28. Albert Einstein (1879–1955): "Religion and Science," *New York Times Magazine*, November 9, 1930, pp. 1–4. Reproduced in part by Alice Calaprice (1941–) in *The Expanded Quotable Einstein*. Princeton, New Jersey: Princeton University Press 2000, p. 207.

29. Albert Einstein (1879–1955): *The World As I See It.* New York: Philosophical Library 1949, p. 7.

30. Albert Einstein (1879–1955): "What I Believe," *Forum and Century* **84**, 193–194 (1930), Reproduced by Alice Calaprice (1941–) *Ibid.* p. 11.

31. The five famous papers of the miracle year 1905 have been discussed by John S. Rigden (1934–2017): *Einstein 1905: The Standard of Greatness.* Cambridge, Massachusetts: Harvard University Press 2005. Also see MBARK, pp. 295–301.

32. Urbain Jean Joseph Le Verrier (1811–1877): "Theorie du movement de Mercure," *Annales de l'Observatoire imperial de Paris, Memoires,* t. 5, Paris: Mallet-Bachelier 1859.

33. Albert Einstein (1879–1955): "Erklärung der Perihelbewegung des Merkur aus der allgemeinen Relativitätstheorie" ("Explanation of the Perihelion Motion of Mercury by Means of the General Theory of Relativity"), *Sitzungsberichte der Preussischen Akademie der Wissenschaften zu Berlin* **11**, 831–839 (1915). [English translation in LGSB, pp. 820–825.]

34. Einstein's conversations or letters quoted by Abraham Pais (1918–2000): *Subtle is the Lord: The Science and the Life of Albert Einstein.* Oxford: Oxford University Press 1982, p. 253.

35. In 1704, Isaac Newton (1642–1727) first speculated that the gravity of massive bodies might bend light, and in 1780 Henry Cavendish (1731–1810) calculated the amount of the Sun's deflection using the Newtonian theory of gravitation. Einstein confirmed these results in 1911, but in his 1915 paper explaining the motion of Mercury, he corrected the calculation by taking the curvature of space-time into account, increasing the light bending by a factor of two to 1.75 seconds of arc for a light ray grazing the Sun's edge.

36. Arthur Stanley Eddington (1882–1944): "On the Future of International Science," *Observatory* **39** (501), 270–272 (June 1916).

37. Willem de Sitter (1872–1934): "On Einstein's Theory of Gravitation, and its Astronomical Consequences," *Monthly Notices of the Royal Astronomical Society* **76**, 699–728 (1916), **77**, 155–184 (1916), **78**, 3–28 (1917).

38. Frank Watson Dyson (1868–1939): "On the Opportunity Afforded by the Eclipse of 1919 May 29 of Verifying Einstein's Theory of Gravitation," *Monthly Notices of the Royal Astronomical Society* **77**, 445 (1917).

39. Alice (Allie) Vibert Douglas (1894–1988): *The Life of Arthur Stanley Eddington.* London: Thomas Nelson and Sons 1956, p. 40.

40. "Joint Eclipse Meeting of the Royal Society and the Royal Astronomical Society," *Observatory* **545**, 389–398 (1919), Frank Watson Dyson (1868–1939), Arthur Stanley Eddington (1882–1944), and Charles Davidson (1875–1970): "A

Determination of the Deflection of Light by the Sun's Gravitational Field, from Observations Made at the Total Eclipse of May 29, 1919," *Philosophical Transactions of the Royal Society (London)* **220**, 291–333 (1920). Reproduced in LGSB, pp. 826–832, and MBARK, pp. 302–312.

41. Russell Alan Hulse (1950–) and Joseph H. Taylor, Jr. (1941–): "A High Sensitivity Pulsar Survey," *Astrophysical Journal Letters* **191**, L59–L61 (1974); "Discovery of a Pulsar in a Binary System," *Astrophysical Journal Letters* **195**, L51–L53 (1975). Reproduced in MBARK, pp. 546–551.

42. Joseph H. Taylor, Jr. (1941–), L. A. Fowler (–), and Peter M. McCulloch (–): "Measurements of General Relativistic Effects in the Binary Pulsar PSR 1913 + 16," *Nature* **277**, 437–440 (1979), Reproduced in MBARK, pp. 551–554; Joseph H. Taylor, Jr. (1941–) and Joel M. Weisberg (–): "A New Test of General Relativity — Gravitational Radiation and the Binary Pulsar PSR 1913+16," *The Astrophysical Journal* **253**, 908–920 (1982); Thibault Damour (1951–) and Joseph H. Taylor, Jr. (1941–): "On the Orbital Period Change of the Binary Pulsar PS 1913+16," *Astrophysical Journal* **366**, 501–511 (1991). Albert Einstein had predicted gravitational radiation in 1916 as a consequence of his recently derived *General Theory of Relativity*. He showed that the gravity waves would travel at the speed of light, as electromagnetic radiation does; but while electromagnetic waves move through space, gravity waves squeeze and stretch space-time itself. The vibrations are so weak, and their interaction with matter so feeble, that Einstein questioned whether they would ever be detected.

43. B. P. Abbott (–) *et al.* (LIGO Scientific Collaboration and Virgo Collaboration): "Observation of Gravitational Waves from a Binary Black Hole Merger," *Physical Review Letters* **116**, 061102, 1–167 (2016). B. P. Abbott (–) *et al.* (with 100 co-authors): "Astrophysical Implications of the Binary Black Hole Merger GW150914," *The Astrophysical Journal Letters* **818**, L22-L37 (2016). B. P. Abbott (–) *et al.* (LIGO Scientific Collaboration and Virgo Collaboration): "GW151226: Observation of Gravitational Waves from a 22-Solar Mass Binary Black Hole Coalescence," *Physical Review Letters* **116**, 241103,1–10. Also see Clara Moskowitz (–): "Gravitational Waves Discovered from Colliding Black Holes," *Scientific American*, February 11, 2016. Dennis Overbye (1944–): "Gravitational Waves Detected, Confirming Einstein's Theory," *New York Times*, February 12, 2016. David Castelvecchi (–): "Gravitational Waves: 6 Cosmic Questions They Can Tackle," *Nature*, February 10, 2016.

44. Stephen W. Hawking (1942–): "Gravitational Radiation from Colliding Black Holes," *Physical Review Letters* **26**, 1344–1346 (1971).

45. Albert Einstein (1879–1955): *Letter to Queen Elizabeth of Belgium* on March 20, 1936. In Helen Dukas (1896–1982) and Banesh Hoffman (1906–1986) (Eds.): *Albert Einstein The Human Side.* Princeton, New Jersey: Princeton University Press 1979, pp. 51–52.

3. Motion within Matter

1. William Blake (1757–1827): *Auguries of Innocence*, written in 1803, first published in Alexander Gilchrist (1828–1861): *The Life of William Blake.* London and Cambridge: Macmillan and Co. 1863. In *The Complete Poetry and Prose of William Blake* [Ed. David V. Erdman (1911–2001): Berkeley: University of California Press 1982, p. 490, lines 1 to 4.

2. Isaac Newton (1642–1727): *Opticks: or, A Treatise of the Reflexions, Refractions, Inflections and Colours of Light.* London: Smith and Walford 1704, New York: Dover 1952, 2012, *Book Three, Part I*, p. 400.

3. John Dalton (1766–1844): "On the Absorption of Gases by Water and Other Liquids," *Memoirs of the Literary and Philosophical Society of Manchester* 1803. *A New System of Chemical Philosophy*, London: Strand 1808, 1810, and 1827.

4. Antoine de Saint-Exupéry (1900–1944): *The Little Prince* (English translation by Katherine Woods). New York: Harcourt, Bracc and World 1943, pp. 87, 93.

5. Henri Becquerel (1852–1908): "Sur les radiations émises par phosphorescence (On the Rays Emitted by Phosphorescence)," *Comptes Rendus de l'Académie des Sciences* **122**, 420–421 (1896); "Sur les radiations invisibles émises par les corps phosphorescents (On the Invisible Rays Emitted by Phosphorescent Bodies," **122**, 501–503 (1896); "Sur diverses propriétés des rayons uranique (On the Diverse Properties of Rays from Uranium)," *Comptes Rendus de l'Académie des Science* **123**, 855–858 (1898).

6. Marie Sklodowska Curie (1867–1934): "Rayons emis par les composes de l'uranium et du thorium (Rays Emitted by Compounds of Uranium and Thorium," *Comptes Rendus de l'Académie des Science* **125**, 1101–1103 (1898). Pierre Curie (1859–1906), Mme. Pierre Curie (1867–1934), and Gustave Bémont (1867–1932): "Sur une nouvelle substance fortement radio-active contenue das la pechblende (On a New, Strongly Radio-active Substance Contained in Pitchblende)," *Comptes Rendus de l'Académie des Science* **127**, 1215–1217 (1898). Pierre Curie (1859–1906): "Sur la radioactivité induite et sur lémanation du radium (Induced Radio-activity and the Emanation from Radium)," *Comptes Rendus de l'Académie des Science* **136**, 223–226 (1903).

7. Ernest Rutherford (1871–1937) and Frederick Soddy (1877–1956): "The Cause and Nature of Radioactivity, Part I, Part II" *Philosophical Magazine* **4**: 370–396, 569–585 (1902).

8. Ernest Rutherford (1871–1937) and Thomas Royds (1884–1955): "Spectrum of Radium Emanations," *Philosophical Magazine* **16**, 313–317 (1908), Ernest Rutherford (1871–1937) and Thomas Royds (1884–1955): "The Nature of the α Particle from Radioactive Substances," *Philosophical Magazine* **17**, 281–286 (1909).

9. Ernest Rutherford (1871–1937): "Radium — the Cause of the Earth's Heat," *Harper's Magazine February* 390–396 (1905).

10. George Gamow (1904–1968): "Zur Quantentheorie der Atomzertrümmerung (On the Quantum Theory of the Atomic Nucleus)," *Zeitschrift für Physik* **52**, 510–515 (1928). Also see his *Constitution of Nuclei and Radioactivity*. Oxford: Oxford University Press 1931.

11. Herman Melville (1819–1891): *Moby Dick*. New York: Harper and Brothers 1851, London: Richard Bentley 1851, London: Marshall Cavendish Paperworks 1987; a reproduction of the 1922 edition, p.141.

12. Ernest Rutherford (1871–1937): "Origin of Actinium and Age of the Earth," *Nature* **123**, 313–314 (1929). Claire Patterson (1922–1995): "Age of the Meteorites and the Earth," *Geochimica et Cosmochimica Acta* **10**, 230–237 (1956).

13. James Clerk Maxwell (1831–1879): "Illustration of the Dynamical Theory of Gases: Part I. On the Motion and Collision of Perfectly Elastic Spheres," *Philosophical Magazine* **19**, 19–32 (1860); "Part II. On the Process of Diffusion of two or more Kinds of moving Particles among one another," *Philosophical Magazine* **20**, 21–27 (1860). In the previous year, Maxwell described why Saturn's rings do not fall down into the planet. The rings contain a vast number of small particles moving around Saturn. See James Clerk Maxwell (1831–1879): *On the Stability of the Motion of Saturn's Rings*. London: Macmillan and Co. 1859. The energy of moving atoms and molecules is called *kinetic energy*, after the Greek word *kinesis* meaning "motion" — the word *cinema* has the same root, referring to motion pictures.

14. Robert Brown (1773–1858): "A Brief Account of Microscopical Observations Made on the Particles Contained in the Pollen of Plants," *London and Edinburgh Philosophical Magazine and Journal of Science* **4**, 161–173 (1828). Also see Albert Einstein (1879–1955): "On the Movement of Small Particles Suspended in Stationary Liquids Required by the Molecular-Kinetic Theory of Heat," *Annalen der Physik* **17**, 549–560 (1905); Jean (Jean-Baptiste) Perrin (1870–1942): "Mouvement brownien et réalité moléculaire," *Annales de Chimie et de Physique* **18**, 1–114 (1909); Jean (Jean-Baptiste) Perrin (1870–1942): *Les Atomes*. Paris: Librairie Félix Alcan 1913.

15. Ernest Rutherford (1871–1937): "The Scattering of α and β Particles by Matter and the Structure of the Atom," *Philosophical Magazine* **21**, 669–688 (1911); "The Structure of the Atom," *Philosophical Magazine* **27**, 488–498 (1914).

16. Francis W. Aston (1877–1945): "Physics at the British Association," *Nature* **106**, 357 (1920). Here Aston states that: "The elements may be considered as composed of hydrogen nuclei, or 'protons' as Sir Ernest Rutherford would have us call them." Also see Francis W. Aston (1877–1945): "The Mass Spectra of Chemical Elements," *Philosophical Magazine and Journal of Science* **39**, 611–625 (1920).

17. James Chadwick (1891–1974): "Possible Existence of a Neutron," *Nature* **129**, 312 (1932), "The Existence of a Neutron," *Proceedings of the Royal Society* **A136**, 692–708 (1932).

18. Jonathan Homer Lane (1819–1880): "On the Theoretical Temperature of the Sun; under the Hypothesis of a Gaseous Mass maintaining its Volume by its Internal Heat, and depending on the Laws of Gases as Known to Terrestrial Experiment," *American Journal of Science and Arts* (2nd series) **50**, 57–74 (1870)." Reproduced by A. J. Meadows (1934–): *Early Solar Physics*. Oxford: Pergamon Press 1970, pp. 257–276.

4. How Light Moves Through Space and Interacts with Matter

1. *The Holy Bible, King James Version, The Epistle of Paul the Apostle to the Romans 1:* 20.

2. *The Holy Bible, King James Version, The First Epistle of Paul the Apostle to the Corinthians, I Corinthians 1:* 10.

3. Michael Faraday (1791–1867): "Observations on Mental Education," a lecture given on May 6, 1854 before the Prince Consort and the Royal Institution, reproduced by Faraday in his *Experimental Researches in Chemistry and Physics*. London: Taylor and Francis 1859, p. 491.

4. Hans Christian Ørsted (1777–1851): "Experiments on the Effect of a Current of Electricity on the Magnetic Needle," *Annals of Philosophy* **16**, 273–276 (1820). Also see André-Marie Ampère (1775–1836): "Conclusions d'un Mémoire sur l'Action Mutuelle de deux courans électriques, sur celle qui existe entre un courant électrique et un aimans, et celle de deux aimans l'un sur l'autre," *Journal de Physique* **91**, 76–78 (1820).

5. Michael Faraday (1791–1867): *Experimental Researches in Electricity, Volume 1, Volume 2*. London: Richard and John Edward Taylor 1849.

6. Michael Faraday (1791–1867): "Thoughts on Ray-vibration," *Philosophical Magazine* **28**, 345–350 (1846), p. 350.

7. Lewis Campbell (1830–1908) and William Garnett (1850–1932): *The Life of James Clerk Maxwell, Second Edition*. London: Macmillan and Co. 1884, p. 135.

8. Lewis Campbell (1830–1908) and William Garnett (1850–1932): *Ibid.* p. 225

9. Simon Schaffer (1955–): "James Clerk Maxwell," in Peter Harman (1943–) and Simon Mitton (1946–) (Eds.): *Cambridge Scientific Minds*. Cambridge, England: Cambridge University Press 2002, p. 137.

10. Basil Mahon (1937–): *The Man Who Changed Everything: The Life of James Clerk Maxwell*. Chichester, England: John Wiley and Sons 2004, p. 173.

11. James Clerk Maxwell (1831–1879): "A Dynamical Theory of the Electromagnetic Field," *Philosophical Transactions of the Royal Society of London* 155, 459–512 (1865); also see James Clerk Maxwell (1831–1879): *A Treatise on Electricity and Magnetism*, Oxford: Clarendon Press 1873.

12. Oliver Heaviside (1850–1925): *Electrical Papers, Second Edition, Volumes 1 and 2*, London: Macmillan Co. 1892, 1894. Reproduced in Providence, Rhode Island: Chelsea Publishing 1970. The quotation is from Nancy Forbes (–) and Basil Mahon (1937–): *Faraday, Maxwell and the Electromagnetic Field*. Amherst, New York: Prometheus Books 2014, p. 240.

13. Heinrich Hertz (1857–1894): "On the Action of a Rectilinear Electric Oscillation Upon a Neighboring Circuit," *Annalen der Physik* (1888), English translation of Hertz' book by Daniel Evan Jones (1860–1941): *Electric Waves: Being Researches On the Propagation of Electric Action with Finite Velocity Through Space*. London and York: Macmillan and Co. 1893.

14. Guglielmo Marconi (1874–1937): "Wireless Telegraphic Communication," Nobel Lecture, December 11, 1909.

15. Ole Rømer (1644–1710): "A demonstration concerning the motion of light," *Philosophical Transactions of the Royal Society* 136, 893–894 (June 25, 1677). Reproduced in MBARK, pp. 117–120.

16. Albert Abraham Michelson (1852–1931): "Experimental Determination of the Velocity of Light," *Proceedings of the American Academy of Arts and Sciences* 27, 71–77 (1878), 28, 124–160 (1879), *Nature* 21, 94–96, 120–122, 226 (1879–1880); "Measurement of the Velocity of Light between Mount Wilson and Mount San Antonio," *Astrophysical Journal* 65, 1–14, 14–22 (1927). Also see Robert A. Millikan (1868–1953): "Biographical Memoir of Albert Abraham Michelson," *Biographical Memoirs of the National Academy of Science* 19, 121–146 (1938).

17. Albert Abraham Michelson (1852–1931): *Light Waves and Their Uses*. Chicago: The University of Chicago Press 1903, pp. 1–2; "Form Analysis," *Proceedings of the American Philosophical Society* 45, 110–116 (1906); "On Metallic Coloring in Birds and Insects," *Philosophical Magazine* (6) 21, 554–569 (1911); Also see Dorothy Michelson Livingston (1906–1994): *The Master of Light: A Biography of Albert A. Michelson*. New York: Charles Scribner's Sons 1973, pp. 4, 5, 15, 16.

18. Christiaan Huygens (1629–1695): *Traité de la lumière*. Paris: Gauthier-Villars Editeurs, Libraires du Bureau des Longitudes 1690, English translation, *Treatise*

on Light, by Silvanus P. Thompson (1851–1916), London: Macmillan and Co. 1912; Isaac Newton (1643–1727): *Opticks or, A Treatise of the Reflections, Refractions, Inflections and Colors of Light*, London: Samuel Smith and Benjamin Walford 1704, Query 18; James Clerk Maxwell (1831–1879): "Lecture on the Aether in 1873", in W. D. Niven (1842–1917) (Ed.): *The Scientific Papers of James Clerk Maxwell, Vol. 2*. Cambridge England: Cambridge University Press 1890, Reproduced New York: Dover 1965, p. 322; For a complete history of the aether, see E. T. (Edmond Taylor) Whittaker (1873–1956): *A History of the Theories of Aether and Electricity (First Edition): From the Age of Descartes to the Close of the Nineteenth Century*. Dublin: Longmans, Green and Co. 1910; *History of the Theories of Aether and Electricity (Second Edition): Vol. 1 The Classical Theories, Vol. 2 The Modern Theories 1900–1926*. London: Nelson 1951.

19. Albert A. Michelson (1852–1931): "The Relative Motion of the Earth and the Luminiferous Aether," *American Journal of Science* (3) **22**, 130–129 (1881). Also see: "On the Application of Interference Methods to Astronomical Measurements," *Philosophical Magazine* **30**, 1–21 (1890). Reproduced in LGSB, pp. 1–7.

20. Albert A. Michelson (1852–1931) and Edward W. Morley (1838–1923): "On the Relative Motion of the Earth and the Luminiferous Ether," *American Journal of Science* **34**, 333–345 (1887).

21. Dorothy Michelson Livingston (1906–1994): *The Master of Light: A Biography of Albert A. Michelson*. New York: Charles Scribner's Sons 1973.

22. Albert Einstein (1879–1955): "Zur Elektrodynamik bewegter Körper (On the Electrodynamics of Moving Bodies)," *Annalen der Physik* **17**, 891–921(1905). [English translation in *The Principle of Relativity*. London: Methuen and Company 1923, reproduced New York: Dover Publications 1952, pp. 35–65.]

23. On January 15, 1931 at a dinner given in Einstein's honor at the Athenaeum of the California Institute of Technology. See "Professor Einstein at the California Institute of Technology: Addresses at the Dinner in His Honor," *Science* **73**, 375–381 (1931). Also see Dorothy Michelson Livingston (1906–1994): *The Master of Light: A Biography of Albert A. Michelson*. New York: Charles Scribner's Sons 1973, p. 335.

24. Max Planck (1858–1947): "Religion und Naturwissenshchaft (Religion and Natural Science)," Lecture delivered in May 1937. English translation in Max Planck (1858–1947): *Scientific Autobiography and Other Papers*, Translated from German by Frank Gaynor (1911–), New York: Philosophical Library 1949, pp. 151–187, quotes from pp. 159, 173.

25. Gustav R. Kirchhoff (1824–1887): "On the Simultaneous Emission and Absorption of Rays of the Same Definite Refrangibility," *Philosophical Magazine* **19**, 193–197 (1860).

26. Joseph Stefan (1835–1893): "Über die Beziehung zwischen Wärmestrahlung und der Temperatur (On the Relationship Between Thermal Radiation and Temperature)," *Sitzungsber. Ak. Wiss. Wien. Math, Naturw.* **79**, 391–397 (1879). Joseph Stefan discovered this expression by comparisons to experimental measurements of the Irish physicist John Tyndall (1829–1893).

27. Samuel Pierpont Langley (1834–1906): "Observations on Invisible Heat-Spectra and the Recognition of Hitherto Unmeasured Wave-lengths, Made at the Allegheny Observatory," *Philosophical Magazine* **21**, 394–409 (1886). Isaac Newton had previously used a glass prism to cast sunlight into an elongated band of colors, or wavelengths, which he called a *spectrum*, from the Latin for "spector."

28. Wilhelm Wien (1864–1928): "Eine neue Beziehung der Strahlung schwarzer Körper zum zweiten Hauptsatz der Wärmetheorie (A New Relation Between the Radiation of Black Bodies and the Second Law of Thermodynamics)," *Sitz. Acad. Wiss. Berlin* **1**, 55 (1893); "On the Division of Energy in the Emission-Spectrum of a Black Body," *Philosophical Magazine* **43**, 214 (1894).

29. Max Planck (1858–1947): "Über das Gesetz der Energieverteilung im Normalspectrum (On the Law of Distribution of Energy in the Normal Spectrum)," *Annalen der Physik* **4**, 553–563 (1901); "Zur Theorie der Wärmestrahlung (On the Theory of Thermal Radiation)," *Annalen der Physik* **31**, 758–768 (1910); *The Theory of Heat Radiation.* [English translation by Morton Masius (1883–1979)] 1914, Reproduced New York: Dover 1959.

30. Albert Einstein (1879–1955): "Über einen die Erzeugung und Verwandlung des Lichtes betreffenden heuristischen Gesichtspunkt (On a Heuristic Point of View about the Creation and Conversion of Light)," *Annalen der Physik* **17**, 132–148 (1905).

31. Gilbert N. Lewis (1875–1946): "The Conservation of Photons," *Nature* **118**, 874–875 (1926).

32. William Hyde Wollaston (1766–1828): "A Method of Examining Refractive and Dispersive Powers, by Prismatic Reflection," *Philosophical Transactions of the Royal Society of London* **92**, 365–380 (1802).

33. Joseph Fraunhofer (1787–1826): *Denkschriften der königlichen Akademie der Wissenschaften zu München* (*Memoranda of the Royal Academy of Sciences in Munich*) **5** 193–226 (1817), *Edinburgh Journal of Science* **8**, 93–96 (1828), Reproduced in MBARK, pp. 204–210.

34. Robert William Bunsen (1811–1899): Letter written by Bunsen to the English chemist Henry Enfield Roscoe (1833–1915) in November 1859. Quoted by Roscoe in: *The Life and Experiences of Sir Henry Enfield Roscoe*, London 1906, p. 71. See Gustav Robert Kirchhoff (1824–1877): "On the Chemical Analysis of the Solar Atmosphere," *Philosophical Magazine and Journal of Science* **21**,

185–188 (1861). Reproduced by A. J. Meadows in: *Early Solar Physics*, Pergamon Press, Oxford 1970, pp. 103–106; Gustav Kirchhoff and Robert Bunsen: "Chemical Analysis of Spectrum Observations," *Philosophical Magazine and Journal of Science* **20**, 89–109 (1860), **22**, 329–349, 498–510 (1861). Also see MBARK, pp. 211–217.

5. The Stars are Moving

1. Henry Wadsworth Longfellow (1807–1882): *Evangeline, A Tale of Acadie, Part the First — III* (1847), New York: Maynard, Marrill & Co. 1893, p. 43, lines 4, 5.
2. This apparent change in stellar location is known as *precession*, and it can be detected with the unaided eye in a human lifetime.
3. Edmond Halley (1656–1742): "Considerations on the Change of the Latitudes of Some of the Principal Fixt Stars," *Philosophical Transactions of the Royal Society of London* **30**, 736–738 (1718). Also see Jacques Cassini (1677–1756): "Du Mouvement apparent des Etoiles fixes en Longitude," *Histoire de l'Académie Royale des Sciences de Paris* (1740), *Mémoires* 273–283.
4. Tobias Mayer (1723–1762): "Du motu fixarum proprio commentatio (Comment on the Proper Motion of Stars)," Lecture to the Göttingen Scientific Society on 12 January 1760.
5. William Herschel (1738–1822): "On the Proper Motion of the Sun and Solar System; with an Account of Several Changes That Have Happened Among the Fixed Stars Since the Time of Mr. Flamsteed," *Philosophical Transactions of the Royal Society* **73**, 274–283 (1783), Reproduced in MBARK, pp. 139–144; "On the Direction and Velocity of the Motion of the Sun and Solar System," *Philosophical Transactions of the Royal Society* **95**, 233–256 (1805); "On the Quantity and the Velocity of the Solar Motion," *Philosophical Transactions of the Royal Society* **96**, 205–237 (1806).
6. Jacobus C. Kapteyn (1851–1922): "Star Streaming," *Report of the 75th meeting of the British Association for the Advancement of Science* (South Africa) 236–265 (1905), Reproduced in LGSB, pp. 514–519.
7. Friedrich Wilhelm Bessel (1784–1846): "On the Parallax of 61 Cygni," *Monthly Notices of the Royal Astronomical Society* **4**, 152–161 (1838). Reproduced in MBARK, pp. 153–159.
8. Edmond Halley (1656–1742): *Catalogus Stellarum Australium*. London: R. Hartford 1679.
9. Edmond Halley (1656–1742): "Some Considerations about the Cause of the Universal Deluge, Laid before the Royal Society, on the 12th of December 1694,"

Philosophical Transactions of the Royal Society of London **33**, No. 383, 123–125 (1724). The great flood that destroyed most living things on Earth is described in *The Holy Bible, King James Version, The First Book of Moses, called Genesis* 6: 17.

10. Angus Armitage (1902–1976): *Edmond Halley*. London: Thomas Nelson 1966, p. 123.

11. Alan Cook (1922–2004): *Edmond Halley: Charting the Heavens and the Seas*. Oxford: Clarendon Press 1998, pp. 163, 164.

12. Christian Doppler (1803–1853): "Über das farbige Licht der Doppelsterne und einiger anderer Gestirne des Himmels (On the Colored Light of Some Binary Stars and Some Other Stars of the Heavens)," *Abhandlungen der königlichen. Böhmischen Gesselschaft der Wissenschaften* (*Proceedings of the Royal Bohemian Society of Sciences*) **2**, 465–482 (1842, 1843). Doppler did not refer to distinct wavelengths, and his proposed color changes of the stars are not perceptible.

13. Hermann C. Vogel (1841–1907): "Determination of the Motions in the Line of Sight by Means of Photography," *Monthly Notices of the Royal Astronomical Society* **50**, 239–242 (1890); "On the Spectrographic Method of Determining the Velocity of Stars in the Line of Sight," *Monthly Notices of the Royal Astronomical Society* **52**, 87–96 (1892). Reproduced in MBARK, pp. 228–232.

14. Lewis Boss (1846–1912): *Preliminary General Catalogue of 6188 Stars for the Epoch 1900: Including Those Visible to the Naked Eye and Other Well-determined Stars*. Washington, D.C.: Carnegie Institution of Washington 1910. Boss states that the direction the Sun travels relative to nearby stars, known as the *solar apex* or *apex of the Sun's way*, is toward R.A. 270.5° ± 1.5° and Dec. + 34.3° ± 1.3°. The star Vega has RA = 279° and Dec. + 38.8°.

15. William Wallace Campbell (1862–1938): *Stellar Motions: With Special Reference to Motions Determined by Means of the Spectrograph*. New Haven, Connecticut: Yale University Press 1913. In these Silliman Lectures of 1909–1910, Campbell gave a speed of solar motion as 19.5 ± 0.6 kilometers per second towards R.A. 268.5° ± 2.0° and Dec. + 25.3° ± 1.8°.

16. Publius Ovidius Naso (43 BC–AD 17/18): *Metamorphoseon libri* (*Books of Transformations*, written in 1 AC), *Book the First, The Giants' War*, lines 26 to 29.

17. Galileo Galilei (1564–1642): *Sidereus Nuncius*, Ventis: Apud Thomon Baglionum 1610. English translation by Albert Van Helden (1940–): *The Sidereal Messenger*. Chicago: University of Chicago Press 1989, p. 62.

18. Thomas Wright (1711–1786): *An Original Theory or New Hypothesis of the Universe, Founded upon the Laws of Nature, and Solving by Mathematical*

Principles the General Phenomena of the Visible Creation; and Particularly the Via Lactea (London: H. Chapele 1750). Facsimile reprint edition entitled *An Original Theory or New Hypothesis of the Universe* edited with an introduction by Michael A. Hoskin (1930–), London: Macdonald & Company 1971, New York: Neale Watson Academic Publications 1971. Also see MBARK, pp. 168–174.

19. Immanuel Kant (1724–1804): *Allgemeine Naturgeschichte und Theorie des Himmels* (*Universal Natural History and Theory of the Heavens*), Part One, *On the Systematic Constitution among the Fixed Stars* 1755. English translation by Ian C. Johnston (1938–), Arlington, Virginia: Richer Resources Publications 2008, pp. 30–40. Also see MBARK, pp. 174–181.

20. William Herschel (1738–1822): "On the Construction of the Heavens," *Philosophical Transactions of the Royal Society of London* **75**, 346–352 (1785); Michael A. Hoskin (1930–): *The Construction of the Heavens: William Herschel's Cosmology*. Cambridge, England: Cambridge University Press 2012.

21. American Institute of Physics: *Oral History Transcript of an Interview with Dr. Harlow Shapley* by Charles Weiner (1932–2012) and Helen Wright (1914–1997) on June 8, 1966.

22. Harlow Shapley (1885–1972): "Thermokinetics of Liometopum Apiculatum Mayr," *Proceedings of the National Academy of Sciences* **6** (4), 204–211 (1920); "Note on the Thermokinetics of Dolichoderine Ants," *Proceedings of the National Academy of Sciences* **10** (10), 436–439 (1924).

23. Robert W. Smith (1952–): *The Expanding Universe: Astronomy's 'Great Debate' 1900–1931*. Cambridge, England: Cambridge University Press 1982, pp. 68 and 78; Michael A. Hoskin (1930–): "The 'Great Debate': What Really Happened," *Journal for the History of Astronomy* **7**, 169–182 (1976); Helen Wright (1914–1997): *Explorer of the Universe, a Biography of George Ellery Hale*. New York: Dutton 1966, p. 326.

24. Owen Gingerich (1931–): "How Shapley Came to Harvard, or Snatching the Prize from the Jaws of Debate," *Journal for the History of Astronomy* **19**, 201–207 (1988).

25. Charles A. Whitney (1929–): *The Discovery of Our Galaxy*. New York: Random House 1971, p. 218.

26. Harlow Shapley (1885–1972): *Beyond the Observatory*. New York: Charles Scribner's Sons 1967, p. 123.

27. Harlow Shapley (1885–1972): *Beyond the Observatory*. New York: Charles Scribner's Sons 1967, pp. 47 and 48.

28. Harlow Shapley (1885–1972): *Of Stars and Men: Human Response to An Expanding Universe*. Boston: Beacon Press 1958, pp. 149 and 150.

29. Pope Pius XII (1876–1958): "The Proofs for the Existence of God in the light of Modern Natural Science," Address to the Pontifical Academy of Sciences, November 22, 1951.

30. Henry Norris Russell (1877–1957): *Fate and Freedom*. New Haven, Connecticut: Yale University Press 1927, pp. 67 and 85.

31. Harlow Shapley (1885–1972): "The Orbits of Eighty-Seven Eclipsing Binaries — a Summary," *Astrophysical Journal* **38**, 158–174 (1913).

32. Harlow Shapley (1885–1972): "On the Nature and Cause of Cepheid Variation," *Astrophysical Journal* **40**, 448–465 (1914).

33. Henrietta Leavitt (1868–1921): "1777 Variables in the Magellanic Clouds," *Annals of the Astronomical Observatory of Harvard College* **60**, No. 4 (1908), "Periods of Twenty-five Variable Stars in the Small Magellanic Cloud," *Harvard College Observatory Circular No. 173*, 1–3 (1912), Reproduced in LGSB, pp. 398–400, and MBARK, pp. 383–389.

34. Harlow Shapley (1885–1972): "Studies Based on the Colors and Magnitudes in Stellar Clusters. VI. On the Determination of the Distances of Globular Clusters," *Astrophysical Journal* **48**, 89–124 (1918), pp. 90, 91. Ejnar Hertzsprung had already determined the mean luminosity of nearby Cepheids of reliably known distance five years before Shapley arrived at a similar result. See Ejnar Hertzsprung (1873–1967): "Über die räumliche Verteilung der Veränderlichen vom δ Cephei-Typus," *Astronomische Nachrichten* **196**, 201–208 (1913). Shapley checked and extended his distance estimates using other luminous stars in the globular clusters, which were assumed to have the same intrinsic luminosity as stars of an identical type and known luminosity near the Sun, and he also used the apparent angular diameters of the clusters to infer their distances under the assumption that they all have the same linear diameter.

35. Harlow Shapley (1885–1972): *Letter from Harlow Shapley to Henry Norris Russell* of October 31, 1917, quoted by Robert W. Smith (1952–): *Ibid*, p. 60.

36. Harlow Shapley (1885–1972): "Studies Based on the Colors and Magnitudes in Stellar Clusters VII, The Distances, Distribution in Space, and Dimensions of 69 Globular Clusters," *Astrophysical Journal* **48**, 154–181 (1918). Also see LGSB, pp. 523–534 and MBARK, pp. 390–397.

37. Harlow Shapley (1885–1972): "Globular Clusters and the Structure of the Galactic System," *Publications of the Astronomical Society of the Pacific* **30**, No. 173, 42–54 (1918).

38. Arthur Stanley Eddington (1882–1944): *Stellar Movements and the Structure of the Universe*. London, Macmillan 1914, p. 32. *Letter from A. S. Eddington to Harlow Shapley* on October 24, 1918, quoted by Robert W. Smith (1952–): *Ibid*. p. 67.

39. Jacobus C. Kapteyn (1851–1922): "First Attempt at a Theory of the Arrangement and Motion of the Sidereal System," *Astrophysical Journal* **55**, 302–327 (1922), Reproduced in LGSB, pp. 542–549.

40. Karl G. Jansky (1905–1950): "A Note on the Source of Interstellar Interference," *Proceedings of the Institute of Radio Engineers* **23**, 1158–1163 (1935), Reproduced in LGSB, pp. 30–33, and MBARK, pp. 455–462; Grote Reber (1911–2002): "Cosmic Static," *Astrophysical Journal* **100**, 279–287 (1944), Reproduced in LGSB, pp. 34–35, and MBARK, pp. 462–464.

41. Eric E. Becklin (1940–) and Gerald "Gerry" Neugebauer (1932–2014): "Infrared Observations of the Galactic Center," *Astrophysical Journal* **151**, 145–161 (1968), Reproduced in LGSB, pp. 67–74, and MBARK, pp. 522–527.

42. Bruce Balick (1943–2014) and Robert L. Brown (1943–2014): "Intense Sub-Arcsecond Structure in the Galactic Center," *Astrophysical Journal* **194**, 265–270 (1974).

43. Sheperd Samuel Doeleman (1967–) *et al.*: "Event-horizon-scale Structure in the Supermassive Black Hole Candidate at the Galactic Center," *Nature* **455**, 78–90 (2008).

44. Andrea M. Ghez (1965–): "Measuring Distance and Properties of the Milky Way's Central Supermassive Black Hole with Stellar Orbits," *Astrophysical Journal* **689**, 1044–1062 (2008).

45. Harlow Shapley (1885–1972): "Studies Based on the Colors and Magnitudes in Stellar Clusters XII, Remarks on the Arrangement of the Sidereal Universe," *Astrophysical Journal* **49**, 311–316 (1919), Reproduced in MBARK, pp. 390–397. Also see Harlow Shapley (1885–1972) and Heber D. Curtis (1872–1942): "The Scale of the Universe, Part 1, Part 2" *Bulletin of the National Research Council of the National Academy of Sciences* (Washington, D.C.) **2**, 171–193 (1921), Reproduced in LGSB, pp. 523–534, and Michael A. Hoskin (1930–): "The 'Great Debate': What Really Happened," *Journal for the History of Astronomy* **7**, 169–182 (1976).

6. The Universe is Expanding and Breaking Away

1. John Milton (1608–1674): *Paradise Lost*. London: Samuel Simmons 1667, *The Second Book, The Argument*, lines 890 to 894. New York: Macmillan Pub. Co. 1993 [Ed. Roy Flannagan (–)], p. 197.

2. Charles Messier (1730–1817): Catalogue des Nébuleuses et des Amas d'Étoiles (Catalogue of Nebulae and Star Clusters), *Mémoires de mathématique et de physique, presentés à l'Académie Royale des Sciences* for 1771. Final 1781 version

of 103 objects published in *Connaissance des Temps for 1784*. English translation by Kenneth Glyn Jones (1915–1995): *Messier's Nebulae and Star Clusters*. London: Faber and Faber 1968, *Second Edition*, Cambridge, England: Cambridge University Press, 1991.

3. William Parsons, The Third Earl of Rosse (1800–1867): "Observations on the Nebulae," *Philosophical Transactions of the Royal Society of London* **140**, 499–514 (1850). Reproduced in MBARK, pp. 189–195.

4. James E. Keeler (1857–1900): "The Crossley Reflector of the Lick Observatory," *Astrophysical Journal* **11**, 325–349 (1900), *Publications of the Astronomical Society of the Pacific* **12**, 144–147 (1900). Modern estimates of the total number of galaxies out to a red shift z is 6 billion z^3, or $6 \times 10^9 \, z^3$; see Kenneth R. Lang (1941–): *Essential Astrophysics*. Heidelberg: Springer 2013, pp. 498, 499.

5. Cornelius Easton (1864–1929): "A New Theory for the Milky Way," *Astrophysical Journal* **12**, 136–158 (1900). By using radio telescopes in the 1950s, which see through interstellar dust, astronomers were eventually able to map out the spiral arms of the Milky Way. The radio astronomers tuned into the emission of interstellar hydrogen atoms, at 21-centimeters wavelength. See Jan H. Oort (1900–1972), Frank J. Kerr (1918–2000), and Gart Westerhout (1927–2012): "The Galactic System as a Spiral Nebula," *Monthly Notices of the Royal Astronomical Society* **118**, 379–380 (1958). Reproduced in LGSB, pp. 643–657, and MBARK, pp. 433–439.

6. William Lowell Putnam, III (1924–2014): "An Apology," in *Origins of the Expanding Universe: 1912–1932, Astronomical Society of the Pacific Conference Series, Volume 471* [Eds. Michael J. Way (–) and Deidre Hunter (–)]. San Francisco: Astronomical Society of the Pacific 2013, pp. 259 to 264.

7. Nicholas U. Mayall (1906–1993): "*Edwin Powell Hubble 1889–1953, Biographical Memoir of the National Academy of Sciences*," Washington, D.C.: National Academy of Sciences 1970, p. 177.

8. Gale E. Christianson (1942–2010): *Edwin Hubble: Mariner of the Nebulae*. New York: Farrar, Straus and Giroux 1995.

9. Edwin P. Hubble (1889–1953): "Photographic Investigations of Faint Nebulae," *Publications of the Yerkes Observatory* **4**, 69–85 (1920).

10. Nicholas U. Mayall (1906–1993): *Ibid.* p. 181.

11. Grace Lillian Burke Hubble (1889–1981): Quoted by Gale E. Christianson (1942–2010): *Ibid.* p. 183, from 1980 record in the Edwin P. Hubble Manuscript Collection, Henry Huntington Library, San Marino, California.

12. Grace Lillian Burke Hubble (1889–1981): *Ibid.* p. 183.

13. Pope Pius XII (1876–1958): "The Proofs for the Existence of God in the light of Modern Natural Science," Address to the Pontifical Academy of Sciences, November 22, 1951.

14. Edwin P. Hubble (1889–1953): *The Nature of Science and Other Lectures.* Westport, Connecticut: Greenwood Press 1954, pp. 18, 19, 40, and 41.

15. Allan R. Sandage (1926–2010): Quoted by John Noble Wilford (1933–): "Sizing Up the Cosmos: An Astronomer's Quest," *New York Times,* March 12, 1991.

16. Vesto M. Slipher (1875–1969): "Spectrographic Observations of Nebulae," *Publications of the American Astronomical Society* **3**, 98–100 (1914).

17. Vesto M. Slipher (1875–1969): "A Spectrographic Investigation of Spiral Nebulae," *Proceedings of the American Philosophical Society* **56**, 403–410 (1917). Reproduced in LGSB, pp. 704–707, and MBARK, pp. 418–420.

18. Henry Norris Russell (1877–1957): "The Highest Known Velocity," *Scientific American* **161** (July), 18–19 (1929); Arthur Stanley Eddington (1882–1944): *The Mathematical Theory of Relativity.* Cambridge, England: Cambridge University Press 1923, pp. 161–166.

19. Ejnar Hertzsprung (1873–1967): *Letter to Vesto M. Slipher* on March 14, 1914, Lowell Observatory Archives. The extraordinarily high velocities of recession meant that some of the spiral nebulae could not long remain a part of the Milky Way system. The combined gravitational pull of the entire 100 billion stars in the Milky Way is not enough to retain any spiral nebula moving at speeds in excess of 1,000 kilometers per second.

20. Harlow Shapley (1885–1972) and Heber D. Curtis (1872–1942): "The Scale of the Universe," *Bulletin of the National Research Council for the National Academy of Sciences* **2**, 171–217 (1921). Reproduced in LGSB, pp. 704–707.

21. Edwin P. Hubble's (1889–1953) *Letter to Harlow Shapley* on February 19, 1924 and *Shapley's Letter to Hubble* on February 27, 1924, Widener Library, Harvard University. Quoted in Robert W. Smith (1952–): *The Expanding Universe: Astronomy's 'Great debate', 1900–1931.* Cambridge, England: Cambridge University Press 1982, pp. 114, 119.

22. Edwin P. Hubble (1889–1953): "Cepheids in Spiral Nebulae," *Publications of the American Astronomical Society* **5**, 261–264 (1925), reprinted in *Observatory* **48**, 139–142 (1925), Reproduced in LGSB, pp. 713–715, and MBARK, pp. 407–414.

23. Ernst Öpik (1893–1985): "An Estimate for the Distance of the Andromeda Nebula," *Astrophysical Journal* **455**, 406–410 (1922). Francis G. Pease (1881–1938): "The Rotation and Radial Velocity of the Central Part of the Andromeda Nebula," *Proceedings of the National Academy of Sciences* **4**, 21–24 (1918).

24. Edwin P. Hubble (1889–1953): "A Relation between Distance and Radial Velocity among Extra-Galactic Nebulae," *Proceedings of the National Academy of Sciences* **15**, 168–173 (1929). Reproduced in LGSB, pp. 725–728, and MBARK pp. 421–424.

25. Carl W. Wirtz (1876–1939): "Einiges zur Statistik der Radialbewegungen von Spiralnebeln und Kugelsternhaufen," *Astronomische Nachrichten* **215**, 349–354 (1922); "De Sitter's Kosmologie und die Radialbewungen der Spiralnebel," *Astronomische Nachrichten* **222**, 21–26 (1924).

26. Kunt Lundmark (1889–1958): "The Motions and the Distances of Spiral Nebulae," *Monthly Notices of the Royal Astronomical Society* **85**, 865–894 (1925).

27. Edwin Hubble (1899–1953) and Milton L. Humason (1891–1972): "The Velocity Distance Relation Amongst Extra-Galactic Nebulae," *Astrophysical Journal* **74**, 43–80 (1931), pp. 57, 58.

28. Edwin Hubble (1899–1953): *Letter to Vesto M. Slipher* on March 6, 1953. Lowell Observatory Archives.

29. Harlow Shapley (1885–1972): *Through Rugged Ways of the Stars*. New York: Charles Scribner's Sons 1969, pp. 57, and 58.

30. Willem de Sitter (1872–1934): "On Einstein's Theory of Gravitation, and its Astronomical Consequences," *Monthly Notices of the Royal Astronomical Society* **76**, 699–728 (1916), **77**, 155–184 (1916), **78**, 3–28 (1917). Arthur Stanley Eddington (1882–1944): *The Mathematical Theory of Relativity*. Cambridge England: Cambridge University Press 1923, p. 162.

31. Edwin Hubble (1899–1953): *Letter to Willem de Sitter* on August 21, 1930. Huntington Library. Quoted by Robert W. Smith (1952–): *Ibid.* pp. 151, 183.

32. Howard P. Robertson (1903–1961): "On Relativistic Cosmology," *Philosophical Magazine and Journal of Science* **5** (31), 835–848 (1928); "On the Foundations of Relativistic Cosmology," *Proceedings of the National Academy of Sciences* **15** (11): 822–829 (1929).

33. Helge Kragh (1944–) and Robert W. Smith (1952–): "Who Discovered the Expanding Universe?" *Journal for the History of Astronomy* **41**, 141–162 (2003).

34. Henry Norris Russell (1877–1957): "The Highest Known Velocity," *Scientific American* **161** (July), 18–19 (1929).

35. Edwin Hubble (1899–1953) and Milton L. Humason (1891–1972): "The Velocity Distance Relation Amongst Extra-Galactic Nebulae," *Astrophysical Journal* **74**, 43–80 (1931). Hubble and Humason preferred the designation *extra-galactic nebulae* for the remote spirals, but we now call them *spiral galaxies* or *elliptical galaxies* depending on their shape. The designation *nebulae* is currently reserved for cloudy, gaseous material enveloping individual bright stars.

36. Milton L. Humason (1891–1972): "The Large Apparent Velocities of Extra-Galactic Nebulae," *Astronomical Society of the Pacific Leaflet* **1**, 149–152 (1931), p. 152.

37. Edwin Hubble (1899–1953) and Milton L. Humason (1891–1972): *Ibid.* p. 80.

38. Edwin P. Hubble (1889–1953): "The Law of Red-Shifts," *Monthly Notices of the Royal Astronomical Society* **113**, 658–666 (1953), p. 666; *The Realm of the Nebulae*. New Haven, Connecticut: Yale University Press 1936, p. 6.

39. Blaise Pascal (1623–1662): *Pensées (Thoughts), Part XV, Transition from Knowledge of Man to Knowledge of God* (1660), English translation by A. J. Krailsheimer (1921–2001), New York: Penguin Classics, 1995, pp. 60, 66 (numbers 199, 201 in papers classified by Pascal).

40. Ernest Rutherford (1871–1937): "Origin of Actinium and Age of the Earth," *Nature* **123**, 313–314 (1929); Clair C. Patterson (1922–1995): "Age of the Meteorites and the Earth," *Geochimica et Cosmochimica Acta* **10**, 230–237 (1956).

41. Walter Baade (1893–1960): "The Resolution of Messier 32, NGC 205, and the Central Region of the Andromeda Nebula," *Astrophysical Journal* **100**, 137–136 (1944). Reproduced in LGSB, pp. 425–430, and MBARK, pp. 744–749.

42. Walter Baade (1893–1960): "A Revision of the Extragalactic Distance Scale," *Transactions of the International Astronomical Union* **8**, 397–398 (1952), Reproduced in LGSB, pp. 750–752, and MBARK, pp. 430–432.

43. Allan Sandage (1926–2010): "Current Problems in the Extragalactic Distance Scale," *The Astrophysical Journal (Supplement)* **127**, 513–526 (1958).

44. Robert P. Kirshner (1949–): *The Extravagant Universe*. Princeton, New Jersey: Princeton University Press 2002, p. 150.

45. Wendy L. Freedman (1957–) *et al*: "Final Results from the *Hubble Space Telescope* Key Project to Measure the Hubble Constant," *Astrophysical Journal* **553**, 47–72 (2001).

46. Martin Ryle (1918–1984): "Evidence for the Stellar Origin of Cosmic Rays," *Proceedings of the Physical Society (London)* **A62**, 491–499 (1949). Also see Hannes Alfvén (1908–1995) and Nicolai Herlofson (1935–): "Cosmic Radiation and Radio Stars," *Physical Review* **78**, 616 (1950). Reproduced in LGSB, pp. 779–780.

47. Thomas Gold (1920–2004): "The Origin of Cosmic Radio Noise," *Proceedings of the Conference on Dynamics of Ionized Media*, Department of Physics, University College, London 1951. Reproduced in LGSB, pp. 782–785.

48. Walter Baade (1893–1960) and Rudolph Minkowski (1895–1976): "Identification of the Radio Sources in Cassiopeia, Cygnus A, and Puppis A," *Astrophysical Journal* **119**, 216–214 (1954). Reproduced in LGSB, pp. 786–791.

49. As remembered by Thomas Gold (1920–2004) and quoted by Simon Singh (1964–) in *Big Bang: The Origin of the Universe*. New York, Forth Estate/Harper Collins 2004, p. 414.

50. Martin Ryle (1918–1984): "The Nature of Cosmic Radio Sources," *Proceedings of the Royal Society (London)* **A248**, 289–308 (1958). Reproduced in LGSB, pp. 792–800.

51. Cyril Hazard (–), M. B. Mackey (–), and A. J. Shimmins (–): "Investigation of the Radio Source 3C 273 by the Method of Lunar Occultations," *Nature* **197**, 1037–1039 (1963). Reproduced in LGSB, pp. 803–808.

52. Maarten Schmidt (1929–): "3C 273: A Star-like Object with Large Red-Shift," *Nature* **197**, 1040 (1963). Reproduced in LGSB, p. 808, and MBARK, 503–507.

53. Martin J. Rees (1942–): "Black hole models for active galactic nuclei," *Annual Review of Astronomy and Astrophysics* **22**, 471–506 (1984).

54. Walter Baade (1893–1960) and Fritz Zwicky (1898–1974): "Cosmic Rays From Super-Novae," *Proceedings of the National Academy of Sciences* **20** (5), 259–263 (1934); "On Super-Novae," *Proceedings of the National Academy of Science* **20** (5), 254–259 (1934). Reproduced in LGSB, pp. 469–473 and MBARK, pp. 339–343.

55. Fritz Zwicky (1898–1974): "A super-nova in NGC 4157," *Publications of the Astronomical Society of the Pacific* **49**, 204 (1937).

56. William A. Fowler (1911–1995) and Fred Hoyle (1915–2001): "Nucleosynthesis in Supernovae," *Astrophysical Journal* **132**, 565–590 (1960).

57. Walter Baade (1893–1960): "The Absolute Photographic Magnitude of Supernovae," *Astrophysical Journal* **88**, 285–304 (1938).

58. Walter Baade (1893–1960) and Fritz Zwicky (1898–1974): "Photographic Light-Curves of the Two Supernovae in IC 4182 and NGC 1003," *Astrophysical Journal* **88**, 411–422 (1938). Rudolph Minkowski (1894–1976): "The Spectra of the Supernovae in IC 4182 and NGC 1003," *Astrophysical Journal* **89**, 156–217 (1939). Also see Virginia Trimble (1943–): "Supernovae. Part 1 — the Events," *Reviews of Modern Physics* **54**, 1183–1224 (1952); "Supernovae. Part 2 — the Aftermath," *Reviews of Modern Physics* **55**, 511–564 (1983).

59. Rudolph Minkowski (1894–1976): "Spectra of Supernovae," *Publications of the Astronomical Society of the Pacific* **53**, 224–225 (1941).

60. See Saul Perlmutter (1959–): "Supernovae, Dark Energy, and the Accelerating Universe," *Physics Today* **56** (4), 53–62 (2003).

61. Adam G. Riess (1969–), William H. Press (1948–), and Robert P. Kirshner (1949–): "A Precise Distance Indicator: Type Ia Supernova Multicolor Light-Curve Shapes," *Astrophysical Journal* **473**, 88–109 (1996).

62. Mario Hamuy (1960–) *et al.*: "The Hubble Diagram of the Calán/Tololo Type Ia Supernovae and the Value of H_0," *Astrophysical Journal* **112**, 2398–2407

(1996), with several dedicated, young, international astronomers that included José Maza, Mark Philips, and Nick Suntzeff.

63. Saul Perlmutter (1959–) *et al.*: "Measurements of the Cosmological Parameters Omega and Lamda from the First Seven Supernovae at $z \geq 0.35$," *Astrophysical Journal* **483**, 565–581 (1997); "Discovery of a Supernova Explosion at Half the Age of the Universe," *Nature* **391**, 51 (1998).

64. Adam G. Riess (1969–) *et al.*: "Observational Evidence From Supernovae for an Accelerating Universe and a Cosmological Constant," *Astronomical Journal* **116**, 1009–1038 (1998), Reproduced in MBARK, pp. 608–617.

65. Saul Perlmutter (1959–) *et al.*: "Measurement of Omega and Lambda from 42 High-Redshift Supernovae," *Astrophysical Journal* **517**, 565–586 (1999), Reproduced in MBARK, pp. 618–623.

66. Alan H. Guth (1947–): "Inflationary Universe: A Possible Solution to the Horizon and Flatness Problems," *Physical Review* **D23**, 347–356 (1981).

67. Alan H. Guth (1947–): "Inflation," *Proceedings of the National Academy of Sciences* **90** (11), 4871–4877 (1993).

68. Andrei D. Linde (1948–): "A New Inflationary Universe Scenario: A Possible Solution of the Horizon, Flatness, Homogeneity, Isotropy and Primordial Monopole Problems," *Physics Letters* **B108**, 389–393 (1982).

69. Anna Ijjas (1985–), Paul J. Steinhardt (1952–), and Abraham Loeb(1962–): "Pop Goes the Universe," *Scientific American* **316**, 32–39 (2017).

70. Allan H. Guth (1947–) and David I. Kaiser (1971–): "Inflationary Cosmology: Exploring the Universe from the Smallest to the Largest Scales," *Science* **307**, 884–890 (2005).

71. Allan H. Guth (1947–), David I. Kaiser (1971–), and Yasunori Nomura (1974–): "Inflationary Paradigm after Planck 2013," *Physics Letters* **B733**, 112–119 (2014).

72. Alexander Vilenkin (1949–): "Creation of Universes from Nothing," *Physics Letters* **B117**, No. 1–2, 25–28 (1982).

73. Andrei D. Linde (1948–): "Quantum Creation of the Inflationary Universe," *Letttere Al Nuovo Cimento* **39**, No. 17, 401–405 (1984).

Part II. Nothing Stays the Same

7. Natural History of the Stars

1. Jorge Louis Borges (1899–1986): "The Garden of Forking Paths," in his *Ficciones*, New York: Grove Press 1962, p. 71.

2. With the advent of telescopes, many fainter novae were discovered, and the term *supernova* was adopted to describe the much brighter and longer lasting events

such as Tycho's supernova of 1572 and Kepler's supernova of 1604. Although these supernovae are uncommon within our own Milky Way Galaxy, they can often be seen amongst the billions of other galaxies.

3. John Donne (1572–1631): *Letter to the Countess of Huntington* (1609–10), lines 5 to 8. In *John Donne: The Major Works* [John Carey (Ed., 1934–)]. Oxford: Oxford University Press 2008, p. 198. The disappearance of this new star may have inspired another English poet, John Milton, who in wrote:

> "The Stars with deep amaze
> Stand fixed in steadfast gaze ...
> But in their glimmering Orbs did glow
> Until their Lord himself bespoke, and bid them go."

See John Milton (1608–1674): *On the Morning of Christ's Nativity, The Hymn* (1629), *Part VI*, lines 68, 69 and 75, 76. John T. Shawcross (Ed., 1924–2011): *The Complete Poetry of John Milton*. New York: Doubleday 1963, p. 66.

4. Tycho Brahe (1546–1601): *De Nova et Nullius Aevi Memoria Prius Visa Stella*, or *On the New and Never Previously Seen Star*. Copenhagen: Haunie 1573, English translation in MBARK, pp. 61–66; Galileo Galilei (1564–1642): Lectures in November 1604 at the University of Padua, Italy. See MBARK, pp. 61–66. Described by Dava Sobel (1947–): *Galileo's Daughter: A Historical Memoir of Science, Faith, and Love*. New York: Walker & Company 1999, pp. 52–53, and by David Freedberg (1948–): *The Eye of the Lynx: Galileo, his Friends, and the Beginning of Modern Natural History*. Chicago: University of Chicago Press 2002, pp. 84–85. By comparing his nightly observations with those of other astronomers in distant lands, Galileo found that the nova displayed no parallax. In contrast, when the Moon is observed from two widely separated places on the surface of the Earth, it shows a small parallax shift in position against the background of the distant stars. Johannes Kepler (1571–1630): *De Stella Nova*. Prague: P. Sessli 1606.

5. Angus Armitage (1902–1976): *William Herschel*. London: Thomas Nelson 1962, New York: Doubleday 1963, p. 21.

6. William Herschel (1738–1822): "Account of a Comet," *Philosophical Transactions of the Royal Society* **71**, 492–501 (1781). Reproduced in MBARK, pp. 128–131.

7. Owen Gingerich (1930–): "The 1784 Autobiography of William Herschel," in his *The Great Copernicus Chase and Other Adventures in Astronomical History*. Cambridge, England: Cambridge University Press 1992, p. 163.

8. William Herschel (1738–1822): "A Letter from William Herschel," *Philosophical Transactions of the Royal Society* **73**, 1–3 (1783). Reproduced in MBARK, pp. 132–133.

9. John Keats (1795–1821): *On First Looking into Chapman's Homer* (1816), lines 9–14. Richard Holmes notes that Balboa, not Cortez, reached the Pacific, and that Keats later replaced "wondering eyes" with "eagle eyes," see Richard Holmes (1945–): *The Age of Wonder: How the Romantic Generation Discovered the Beauty and Terror of Science.* New York: Vintage Books, Random House 2008, p. 207.

10. William Herschel (1738–1822): "Catalogue of One Thousand new Nebulae and Clusters of Stars," *Philosophical Transactions of the Royal Society of London* **76**, 457–499 (1786), "Catalogue of a second Thousand of new Nebulae and Clusters of Stars; with a few introductory Remarks on the Construction of the Heavens," *Philosophical Transactions of the Royal Society of London* **79**, 212–255 (1789), "Catalogue of 500 new Nebulae, nebulous Stars, planetary Nebulae and Clusters of Stars; with Remarks on the Construction of the Heavens," *Philosophical Transactions of the Royal Society of London* **92**, 477–528 (1802). Also see Michael A. Hoskin (1930–): *William Herschel and the Construction of the Heavens.* London: Oldbourne 1963. *The Construction of the Heavens: William Herschel's Cosmology.* Cambridge, England: Cambridge University Press 2012. During this time, William also discovered infrared radiation when he put sunlight through a prism to spread it into its colors and used a thermometer to measure their temperatures. See William Herschel (1738–1822): "Experiments on the Refrangibility of the Invisible Rays of the Sun," *Philosophical Transactions of the Royal Society of London* **90**, 284–292 (1800). Herschel noticed that an unseen portion of the sunlight produced heat beyond the red edge of the visible spectrum. Because the thermometer recorded higher temperatures in this invisible sunlight than in normal visible sunlight, Herschel called them calorific, or heat, rays. The term infrared, referring to wavelength rather than heat, did not appear until the late 19th century.

11. Charles Messier (1730–1817): Catalogue des Nébuleuses et des Amas d'Étoiles (Catalogue of Nebulae and Star Clusters). *Mémoires de mathématique et de physique,* prepared in 1781, in *Connaissance des Temps for 1784.* English translation by Kenneth Glyn Jones (1915–1995): *Messier's Nebulae and Star Clusters.* London: Faber and Faber 1968.

12. William Herschel (1738–1822): "Catalogue of a Second Thousand of new Nebulae and Clusters of Stars with a few introductory Remarks on the Construction of the Heavens," *Philosophical Transactions of the Royal Society* **79**, 212–255 (1789), p. 226, *Introductory Remarks,* pp. 212–226. Quoted by Angus Armitage (1902–1976): *William Herschel.* London: Thomas Nelson 1962, New York: Doubleday 1963, p. 118.

13. Erasmus Darwin (1731–1802): *The Botanic Garden, Part 1, The Economy of Vegetation.* London: J. Johnson 1791, *Canto I, Part I,* lines 97–114.

Erasmus' grandson, Charles Darwin, would continue these interests in his *The Origin of the Species by Means of Natural Selection*. London: John Murray 1859.

14. Erasmus Darwin (1731–1802): *The Botanic Garden, Part 1, The Economy of Vegetation*. London: J. Johnson 1791, *Canto IV*, lines 371–388, and *Canto IV, Part X*, lines 9–31.

15. William Herschel (1738–1822): "On Nebulous Stars, Properly So Called," *Philosophical Transactions of the Royal Society of London* **81**, 71–88 (1791). Quoted by Angus Armitage (1902–1976): *William Herschel*. London: Thomas Nelson 1962, New York: Doubleday 1963, p. 114. This object is now known as the planetary nebula NGC 1514. Herschel named such an object a *planetary nebula*, since it seemed to have both the round disk of planets and the pale light of nebulae. In his time, a planetary nebula seemed to be a pre-stellar object about to gather into a nebulous star and from thence to a true star. Modern telescopes of larger size have revealed that planetary nebulae, including those observed by Herschel, mark the end of some stars' lives, rather than their beginning. They display an outer expanding atmosphere that has been expelled by the gusty winds of a dying central star.

16. William Herschel (1738–1822): "Observations Relating to the Sidereal Part of the Heavens," *Philosophical Transactions of the Royal Society of London* **104**, 248–284 (1814), p. 284. Quoted by Michael A. Hoskin (1930–): *The Construction of the Heavens: William Herschel's Cosmology*. Cambridge, England: Cambridge University Press 2012, p. 162.

17. William Herschel (1738–1822): "On the Utility of Speculative Inquiries," read on April 14, 1780 to the Philosophical Society of Bath. Quoted by Stanley L. Jaki (1924–2009): *The Road of Science and the Ways of God*. Chicago: The University of Chicago Press 1978, pp. 110, 111. Also see Michael A. Hoskin (1930–): "William Herschel and God," *Journal for the History of Astronomy* **14**, 1–6 (2014). William has been extensively misquoted as saying "an undevout astronomer must be mad" — which the English poet Edward Young wrote in his *Night Thoughts*. Edward Young (1683–1765): *The Complaint: or Night Thoughts on Life, Death and Immortality*. London: R. Dodsley 1743, *Part IX*, lines 771–773.

18. Constance A. Lubbock (1905–1993): *The Herschel Chronicle*. Cambridge, England: Cambridge University Press 1933, p. 197. Quoted by Angus Armitage (1902–1976): *William Herschel*. London: Thomas Nelson 1962, New York: Doubleday 1963, p. 35.

19. Joseph Addison (1672–1719): *Hymn, The Spacious Firmament on High* or *Creation* (1712), lines 1 to 4. See Arthur Quiller-Couch (Ed., 1863–1944): *The*

Oxford Book of English Verse: 1250–1900, Oxford: Oxford University Press 1900, *No. 433 Hymn*, p. 496.

20. Joseph Haydn (1732–1809): *The Creation*, composed 1796 to 1798; Ludwig van Beethoven's (1770–1827): *Symphony No. 9 in D minor, Op. 125, Fourth Movement, Ode to Joy*, composed 1822 to 1824; musical setting of parts of the German poet Friedrich von Schiller's (1759–1805) poem, published in 1786. This final movement to the *Ninth Symphony* was used by Leonard Bernstein (1918–1990) in a concert to celebrate the fall of the Berlin Wall and has been adopted as the anthem of the European Union.

8. How the Sun and Planets Came into Being

1. Alexander Pope (1688–1744): *Essay on Man. The First Epistle*, 1732, lines 23 to 28. Quoted by Immanuel Kant (1724–1804): *Universal Natural History and Theory of the Heavens, Part Three*, 1755. English translation by Ian C. Johnston (1938–), Arlington, Virginia: Richer Resources Publications 2008, p. 130.

2. Immanuel Kant (1724–1804): *Allgemeine Naturgeschichte und Theorie des Himmels* (*Universal Natural History and Theory of the Heavens*), *Part One, On the Systematic Arrangement of the Fixed Stars. Part Two, Section One, Concerning the Origin of the Planetary World Structure in General and the Causes of its Movements*, 1755. English translation by Ian C. Johnston (1938–), Arlington, Virginia: Richer Resources Publications 2008, p. 22–49.

3. Immanuel Kant (1724–1804): *Allgemeine Naturgeschichte und Theorie des Himmels* (*Universal Natural History and Theory of the Heavens*): *Of the Creation of the Whole Extent of its infinitude in Space as Well as Time*, 1755. English translation by William Hastie (1842–1903) and reproduced in Milton K. Munitz (1913–1995): *Theories of the Universe: From Babylonian Myth to Modern Science*. New York: The Free Press 1957, p. 238.

4. Immanuel Kant (1724–1804): *Ibid.* p. 246.

5. Nicolas de Caritat, Marquis de Condorcet (1743–1794): Quoted by Charles Coulston Gillispie (1918–2015): *Pierre-Simon Laplace 1749–1827: A Life in Exact Science*. Princeton, New Jersey: Princeton University Press 1997, p. 5.

6. Immanuel Kant (1724–1804): *Allgemeine Naturgeschichte und Theorie des Himmels* (*Universal Natural History and Theory of the Heavens*), *Part One, On the Systematic Arrangement of the Fixed Stars*, 1755. English translation by Ian C. Johnston (1938–), Arlington, Virginia: Richer Resources Publications 2008, p. 42.

7. Immanuel Kant (1724–1804): *Ibid.* p. 47.

8. Stanley L. Jaki (1924–2009): "The Five Forms of Laplace's Cosmogony," *American Journal of Physics* **44**, 4–11 (1976).

9. Napoléon Bonaparte (1769–1821): *Correspondence de Napoléon Ier* **30** (1870), p. 330. Quoted by Charles Coulston Gillispie (1918–2015): *Pierre-Simon Laplace 1749–1827: A Life in Exact Science.* Princeton, New Jersey: Princeton University Press 1997, p. 176.

10. Napoléon Bonaparte (1769–1821): Quoted by Roger Hahn (1932–2011) in his *Pierre-Simon Laplace 1749–1827: A Determined Scientist.* Cambridge Massachusetts: Harvard University Press 2008, p. 191.

11. Pierre-Simon Laplace (1749–1827): Quoted by Michael J. Crowe (1936–), *The Extraterrestrial Life Debate 1750–1900.* Cambridge, England: Cambridge University Press 1986, p. 78. From Hervé Faye (1814–1902), *Sur l'origine du monde*, second edition, Paris: Gauthier-Villars 1885, p. 131.

12. William Herschel (1738–1822): Quoted by Constance A. Lubbock (1855–1939), *The Herschel Chronicle*, Cambridge, England: Cambridge University Press 1933, p. 310. Also by Michael Hoskin (1930–): "William Herschel and God," *Journal of the History of Astronomy* **45**, 247–252 (2014).

13. Michel Mayor (1942–) and Pierre-Yves Frei (1964–): *New Worlds in the Cosmos: The Discovery of Exoplanets.* Cambridge, England: Cambridge University Press 2003, p. 18.

14. Michel Mayor (1942–) and Didier Queloz (1966–): "A Jupiter-Mass Companion to a Solar-Type Star," *Nature* **378**, 355–359 (1995). Reproduced in MBARK, pp. 598–603.

15. Geoffrey W. Marcy (1954–) and R. Paul Butler (1960–): "A planetary companion to 70 Virginis," *Astrophysical Journal Letters* **464**, L147–L151 (1996). R. Paul Butler (1960–) and Geoffrey W. Marcy (1954–): "A Planet Orbiting 47 Ursae Majoris, *Astrophysical Journal Letters* **464**, L153–L156 (1996). Reproduced in MBARK, pp. 603–607.

16. Michael J. Crowe (1936–): *The Extraterrestrial Life Debate 1750–1900: The idea of a plurality of worlds from Kant to Lowell.* Cambridge, England: Cambridge University Press 1986; Steven J. Dick (1949–): *Plurality of Worlds: The Origins of the Extraterrestrial Life Debate from Democritus to Kant.* Cambridge, England: Cambridge University Press 1982.

17. Immanuel Kant (1724–1804): *Universal Natural History and Theory of the Heavens, Part Three*, 1755. English translation by Ian C. Johnston (1938–), Arlington, Virginia: Richer Resources Publications 2008, p. 138. Also in Stanley L. Jaki (1924–2009): "An English Translation of the Third Part of Kant's *Universal Natural History and Theory of the Heavens*" in *Cosmology, History of Science, and Theology*, Eds. Allen D. Breck (1914–2000) and W. Yourgrau (1908–1974), New York: Plenum Press 1977, p. 396.

18. Pierre-Simon Laplace (1749–1827): *Exposition du Système du Monde*. Paris: Bachelier 1824. English translation by Henry H. Harte (1790–1848): *The System of the World*, Dublin: University Press 1830, *Volume II, Book V, Chapter VI*, p. 326. Also see MBARK, pp. 121–123.

19. Arthur Stanley Eddington (1882–1944): "Herschel's Researches on the Structure of the Heavens," *Occasional Notes of the Royal Astronomical Society* 1, 27–32 (1938–1941), p. 30. Quoted by Michael A. Hoskin (1930–): *Discoverers of the Universe: William and Caroline Herschel*. Princeton, New Jersey: Princeton University Press 2011, p. 74.

9. The Ways Stars Shine

1. John Updike (1932–2009): "The Astronomers," in *Pigeon Feathers and Other Stories*. New York: Alfred A. Knopf 1963, p. 186.

2. Cecilia H. Payne (1900–1979): "The Relative Abundances of the Elements," in her *Stellar Atmospheres*. Cambridge, Massachusetts: Harvard University Press 1925. Reproduced in LGSB, pp. 243–253, and MBARK, pp. 250–256.

3. Bengt Strömgren (1908–1987): "The Opacity of Stellar Matter and the Hydrogen Content of the Stars," *Zeitschrift für Astrophysik* 4, 118–152 (1932).

4. Pierre Jules César (P. J. C. Janssen) (1824–1907): "Éclipse de soleil du 18 Aout 1868," *Annales de Chimie et de Physique* 15, 414–426 (1878). *Summary of some of the Results Obtained at Cocanada, during the Eclipse last August*. Letter to the French Academy of Sciences.

5. Joseph Norman Lockyer (1836–1920): "Spectroscopic Observations of the Sun. III, IV," *Proceedings of the Royal Society* 17, 350–356, 415–418 (1869); *The Chemistry of the Sun*. London: Macmillan and Co. 1887. Reproduced in MBARK, pp. 268–271.

6. William Ramsay (1852–1916): "Helium, a Gaseous Constituent of Certain Minerals. Part I, II," *Proceedings of the Royal Society of London* 58, 80–89, 59, 325–330 (1895). Helium is one of the noble elements that also include neon, argon, krypton, xenon and radon. These so-called *noble elements* do not combine with most other chemical elements, behaving like people of nobility who are unwilling to associate with ordinary, common folks. Ramsay received the Nobel Prize in Chemistry in 1904 for his discovery of these inert gaseous elements in the Earth's atmosphere.

7. For seven important papers by A. S. Eddington on various topics see LGSB.

8. Arthur Stanley Eddington (1882–1944): *The Internal Constitution of the Stars*. Cambridge, England: Cambridge University Press 1926. Also see Kenneth R.

Lang (1941–): *The Life and Death of Stars.* Cambridge, England: Cambridge University Press 2013.

9. Arthur Stanley Eddington (1882–1944): *Stars and Atoms.* London: Oxford University Press 1927; *The Nature of the Physical World: The Gifford Lectures 1927.* Cambridge, England: Cambridge University Press 1928; *Science and the Unseen World: Swarthmore Lecture 1929.* New York: Macmillan 1929; *The Expanding Universe.* Cambridge, England: Cambridge University Press 1933; *New Pathways in Science.* Cambridge, England: Cambridge University Press 1935.

10. Matthew Stanley (1975–): *Practical Mystic: Religion, Science and A. S. Eddington.* Chicago: University of Chicago Press 2007.

11. Arthur Stanley Eddington (1882–1944): *Science and the Unseen World.* New York: Macmillan 1929, pp. 23, 41, and 90.

12. Arthur Stanley Eddington (1882–1944) and Emanuel Haldeman-Julius (1889–1951): *Why I Believe In God: Science and Religion: As a Scientist Sees It.* Girard, Kansas: Haldeman-Julius Publications 1930, p. 7. *Science and the Unseen World.* New York: Macmillan 1929, p. 42.

13. Arthur Stanley Eddington (1882–1944): *New Pathways in Science.* Cambridge, England: Cambridge University Press 1935, p. 317.

14. J. W. N. Sullivan (1886–1937): "Interview with Sir A. S. Eddington," in *Contemporary Mind: Some Modern Answers.* London: Humphrey Toulmin 1934, pp. 124–129; quotation p. 125.

15. Arthur Stanley Eddington (1882–1944) and Emanuel Haldeman-Julius (1889–1951): *Why I Believe In God: Science and Religion: As a Scientist Sees It.* Girard, Kansas: Haldeman-Julius Publications 1930, p. 11, *Science and the Unseen World.* New York: Macmillan 1929, pp. 43 and 53.

16. Arthur Stanley Eddington (1882–1944): "The Internal Constitution of the Stars," *Nature* **106**, 14–20 (1920), pp. 357, 358, *The Observatory* **43**, 341–358 (1920), Reproduced in LGSB, pp. 281–290, see p. 289.

17. Jonathan Homer Lane (1819–1880): "On the Theoretical Temperature of the Sun; under the Hypothesis of a Gaseous Mass maintaining its Volume by its Internal Heat, and depending on the Laws of Gases as Known to Terrestrial Experiment," *American Journal of Science and Arts* (2nd series) **50**, 57–74 (1870).

18. William Thomson (1824–1907): "On the Age of the Sun's Heat," *Macmillan's Magazine*, March 5, 288–293 (1862), *Popular Lectures* **I**, 349–368. William Thomson is better known today as Lord Kelvin.

19. Alice Vibert Douglas (1894–1988): *The Life of Arthur Stanley Eddington.* London: Thomas Nelson and Sons 1956, p. 60. Also see LGSB, p. 287.

20. Arthur Stanley Eddington (1882–1944): *Ibid.* See LGSB, pp. 287–288.

21. Albert Einstein (1879–1955): "The Principle of the Conservation of the Motion of the Center of Gravity," *Annalen der Physik* **20**, 627–633 (1906). [English translation in LGSB, pp. 276–280.]

22. Arthur Stanley Eddington (1882–1944): *Ibid.* See LGSB p. 288.

23. Hans Bethe (1906–2005): "Energy Production in Stars," *Physical Review* **55**, 434–456 (1939). Reproduced in LGSB, pp. 320–338, and MBARK, pp. 349–357.

24. Carl Friedrich von Weizsäcker (1912–2007): "Element Transformation Inside Stars II," *Physikalische Zeitschrift* **39**, 633–646 (1938). [English translation in LGSB, pp. 309–319.] C. F. von Weizsäcker has written serious, philosophical books that include Christianity and the nature of science; see his *The History of Nature.* Chicago: The University of Chicago Press 1949, and *The Relevance of Science. Creation and Cosmogony, Gifford Lectures 1959–1960.* St. James's Place, London: Collins 1964.

25. Arthur Stanley Eddington (1882–1944): *Stars and Atoms.* London: Oxford University Press 1927, p. 102, *Internal Constitution of the Stars.* Cambridge, England: Cambridge University Press 1926, p. 301.

26. Friedrich Wilhelm Bessel (1784–1846): "On the Parallax of 61 Cygni," *Monthly Notices of the Royal Astronomical Society* **4**, 152–161 (1838). Bessel measured an annual parallax for 61 Cygni of 0.31 seconds of arc, close to the modern value of 0.286. Reproduced in MBARK, pp. 153–159.

27. Arthur Stanley Eddington (1882–1944): *Science and the Unseen World.* New York: Macmillan 1929, p. 14.

28. Arthur Stanley Eddington (1882–1944): "The Radiative Equilibrium of the Sun and Stars," *Monthly Notices of the Royal Astronomical Society* **77**, 16–35, 596–597 (1917). Reproduced in LGSB, pp. 225–235, and MBARK, pp. 258–259.

29. Arthur Stanley Eddington (1882–1944): "On the Relation between the Masses and Luminosities of the Stars," *Monthly Notices of the Royal Astronomical Society* **84**, 308–332 (1924). Reproduced in LGSB, pp. 291–302, and MBARK, pp. 259–260.

30. James Hopwood Jeans (1877–1946): "Meeting of the Royal Astronomical Society Friday 1925, January 9," *The Observatory* **48**, 28–39 (1925), p. 30.

31. Rupert Brooke (1887–1915): *1914 IV. The Dead*, 1915, lines 9 to 14. Quoted by Eddington in his *The Nature of the Physical World.* Cambridge, England: Cambridge University Press 1928, p. 317.

32. Arthur O'Shaughnessy (1844–1957): *Ode (poem)*, 1874. Quoted by Eddington *ibid*, pp. 325–326.

33. Henry Norris Russell (1877–1957): "Arthur Stanley Eddington 1882–1944," *The Astrophysical Journal* **101**, 133–135 (1945).

10. The Paths of Stellar Life

1. Robert Frost (1874–1963): *The Road Not Taken* (1916). In Edward Connery Lathem (Ed., 1926–2009): *The Poetry of Robert Frost*. New York: Holt Rinehart and Winston 1969, p. 105.

2. Henry Norris Russell (1877–1957): "Spiritual Autobiography," in Louis Finkelstein (1895–1991) (Ed.): *Thirteen Americans: Their Spiritual Autobiographies*. Port Washington, New York: Kennikat Press 1953, p. 33.

3. Henry Norris Russell (1877–1957): *Ibid.* p. 33.

4. Charles A. Young (1834–1908): *God's Glory in the Heavens*. Princeton Theological Seminary: George W. Burroughs Printer 1894, pp. 5, 12, 13, and 17.

5. *The Holy Bible, King James Version, The Book of Psalms, Psalm 19*: 1.

6. Henry Norris Russell (1877–1957): *Letter to Woodrow Wilson* on April 6, 1905, In Arthur S. Link (1920–1998) *et al.* (Eds.) *The Papers of Woodrow Wilson, Volume 16: 1905–1907*, p. 81. Quoted by David H. DeVorkin (1944–): *Henry Norris Russell: Dean of American Astronomers*. Princeton, New Jersey: Princeton University Press 2000, p. 69.

7. Henry Norris Russell (1877–1957): *Fate and Freedom*. New Haven, Connecticut: Yale University Press 1927, p. 67. Russell italicized the words *How* and *Why*.

8. Henry Norris Russell (1877–1957): "Spiritual Autobiography," in Louis Finkelstein (1895–1991) (Ed.): *Thirteen Americans: Their Spiritual Autobiographies*, Port Washington, New York: Kennikat Press 1953, pp. 34, 47.

9. Henry Norris Russell (1877–1957): *Fate and Freedom*. New Haven, Connecticut: Yale University Press 1927, pp. 70, 71, 79, 80.

10. Henry Norris Russell (1877–1957): *Ibid.* p. 85. Russell italicized the word *why*.

11. Henry Norris Russell (1877–1957): *Ibid.* pp. 151, 154.

12. Harlow Shapley (1885–1972): "Henry Norris Russell 1877–1957," *Biographic Memoirs of the National Academy of Science*. Washington, D.C.: National Academy of Science 1958.

13. David H. DeVorkin (1944–): *Henry Norris Russell: Dean of American Astronomers*. Princeton, New Jersey: Princeton University Press 2000.

14. Annie Jump Cannon (1863–1941) and Edward C. Pickering (1846–1919): *Annals of the Astronomical Observatory of Harvard College No. 91–99* (1918–1924). Also see MBARK, pp. 233–240.

15. August Ritter (1826–1908): "On the Constitution of Gaseous Celestial Bodies," *Astrophysical Journal* **8**, 293–315 (1898).

16. Norman Lockyer (1836–1920): *Inorganic Evolution as Studied by Spectrum Analysis*. London: Macmillan and Co. 1900. George Ellery Hale (1868–1938):

The Study of Stellar Evolution. Chicago: University of Chicago Press 1908, pp. 1–8, and 186–203.

17. Henry Norris Russell (1877–1957): "Relations between the Spectra and Other Characteristics of Stars," *Popular Astronomy* **22**, 275–294 (1914). Reproduced in LGSB, pp. 212–220, and MBARK, pp. 245–249.

18. Ejnar Hertzsprung (1873–1967): "Zur Strahlung der Sterne" (On the Radiation of Stars), *Zeitschrift für Wissenschaftliche Photographie* **3**, 429–442 (1905). [English translation in LGSB, pp. 208–211, also see MBARK, pp. 243–245]; Ejnar Hertzsprung (1873–1967): "Über die Verwendung Photographischer Effektiver Wellenlaengen zur Bestimmung von Farbenaequivalenten," *Publikationen des Astrophysikalischen Observatoriums zu Potsdam* 22, Bd. 1, Nr. 63 (1911).

19. Henry Norris Russell (1877–1957): "'Giant' and 'Dwarf' Stars," *Observatory* **36**, 324–329 (1913).

20. Owen Gingerich asked Hertzsprung about the invention of the dwarf-giant distinction, and he replied: "I hasten to say that I have avoided the expressions 'giant' and 'dwarf' because the stars are not very different in mass but in density. They are more or less 'swollen'." See Owen Gingerich (1930–): "The Critical Importance of Russell's Diagram," in Eds. Michael J. Way (–) and Deidre Hunger (–): *Origins of the Expanding Universe: 1912–1932. ASP Conference Series. Vol. 471.* San Francisco: Astronomical Society of the Pacific 2013, p. 3.

21. Albert A. Michelson (1852–1931) and Francis G. Pease (1881–1938): "Measurement of the Diameter of Alpha Orionis [Betelgeuse]," *The Astrophysical Journal* **53**, 249–259 (1921).

22. Arthur Stanley Eddington (1882–1944): "On the Relation between the Masses and Luminosities of the Stars," *Monthly Notices of the Royal Astronomical Society* **84**, 308–332 (1924). Reproduced in LGSB, pp. 291–302, and MBARK, pp. 269–260.

23. Henry Norris Russell (1877–1957): "The Problem of Stellar Evolution," *Nature* **116**, 209–212 (1925).

24. Bengt Strömgren (1908–1987): "On the Interpretation of the Hertzsprung-Russell-Diagram," *Zeitschrift für Astrophysik* **7**, 222–269 (1933).

25. Merle F. Walker (1926–): "Studies of Extremely Young Clusters I: NGC 2264," *Astrophysical Journal Supplement* **2**, 365–387 (1956); Chushiro Hayashi (1920–2010): "Stellar Evolution in Early Phases of Gravitational Contraction," *Publications of the Astronomical Society of Japan* **13**, 450–452 (1961). Reproduced in LGSB, pp. 364–373.

26. Ernst J. Öpik (1893–1985): "Stellar Structure, Source of Energy, and Evolution," *Publications de l'Observatoire Astronomique de l'Université de Tartu*

30, No. 3, 1–115 (1938), partially reproduced in LGSB, pp. 342–347, and MBARK, pp. 377–380.

27. Allan R. Sandage (1926–2010): "The Color-Magnitude Diagram for the Globular Cluster M 3," *The Astronomical Journal* **58**, 61–75 (1953).
28. Allan R. Sandage (1926–2010): "Current Problems in the Extragalactic Distance Scale," *The Astrophysical Journal* **127** (3), 513–516 (1958).
29. Allan R. Sandage (1926–2010) and Martin Schwarzschild (1912–1997): "Inhomogeneous Stellar Models II: Models with Exhausted Cores in Gravitational Contraction," *The Astrophysical Journal* **116**, 463–476 (1952), Reproduced in LGSB, pp. 353–363.
30. Spencer R. Weart (1942–): "Oral History Transcript, Dr. Allan Sandage, May 23, 1978." *American Institute of Physics*, p. 6.
31. Allan R. Sandage (1926–2010): "A Scientist Reflects on Religious Belief," *Truth Journal, Leadership University*, 1 July 2002, p. 2.

11. The Ways Stars Die

1. Julian Huxley (1887–1975): *Cosmic Death* (1923), in *The Captive Shrew and Other Poems of a Biologist*, London and New York: Harper and Brothers 1933, p. 30.
2. In 1786, the English astronomer William Herschel provided the designation *planetary nebula* for their round shapes, which resembled the resolved disks of planets rather than unresolved point-like stars. Nearly a century later another Englishman, William Huggins, used his spectroscope to find a trio of emission lines in one of them, the Cat's Eye Nebula. See William Herschel (1738–1822): "Catalogue of One Thousand New Nebulae and Clusters of Stars," *Philosophical Transactions of the Royal Society of London* **76**, 457–499 (1786), p. 492, and William Huggins (1824–1920): "On the Spectra of Some Nebulae," *Philosophical Transactions of the Royal Society of London* **154**, 437–444 (1864). Reproduced in MBARK, pp. 218–228.
3. Ira S. Bowen (1898–1973): "The Origin of the Nebulium Spectrum," *Nature* **120**, 473 (1927); "The Origin of the Nebular Lines and the Structure of the Planetary Nebulae," *The Astrophysical Journal* **67**, 1–15 (1928). Reproduced in LGSB, pp. 581–587.
4. Henry Norris Russell (1877–1957) recalled this discovery of the anomalous spectrum of o Eridani B in a colloquium given at Princeton University Observatory in 1954. Mrs. Fleming was once Pickering's maid. After famously stating that his maid could do a better job than his male assistants, Pickering hired her to work at the Observatory. See LGSB, pp. 430–431, and Eds. A. G. Davis Philip (1929–) and D. H. de Vorkin (1944–): *In Memory of Henry Norris Russell*, Dudley Observatory Report No. 13 (1977), pp. 90–107.

5. Walter S. Adams (1876–1956): "An A-Type Star of Very Low Luminosity," *Publications of the Astronomical Society of the Pacific* **26,** 198 (1914); "The Spectrum of the Companion of Sirius," *Publications of the Astronomical Society of the Pacific* **27**, 236–237. Reproduced in LGSB, pp. 430–432, and MBARK, pp. 327–331.

6. Arthur Stanley Eddington (1882–1944): *Stars and Atoms*, Oxford Clarendon Press 1927, p. 50.

7. Ralph H. Fowler (1889–1944): "On Dense Matter," *Monthly Notices of the Royal Astronomical Society* **87**, 114–122 (1926). Reproduced in LGSB, pp. 433–439, and MBARK, pp. 331–333. This degenerate pressure is unaffected by temperature, and the degeneracy is a mathematical term that describes a limiting situation rather than a person with impaired virtue.

8. Jesse L. Greenstein (1909–2002), J. Beverly Oke (1928–2004), and Harry L. Shipman (–): "Effective Temperature, Radius, and Gravitational Redshift of Sirius B," *Astrophysical Journal* **169**, 563–566 (1971); Jesse L. Greenstein (1909–2002) and Virginia L. Trimble (1943–): "The Gravitational Redshift of 40 Eridani B," *Astrophysical Journal Letters* **175**, L1–L5 (1972).

9. Lord (George Gordon) Byron (1788–1824): *Darkness* (1816). London: John Murray 1816.

10. Wilhelm Anderson (1880–1940): "Über die Grenzdichte der Materie und der Energie (About the Interface of Matter and Energy)," *Zeitschrift für Physik* **56**, 851–856 (1929); Edmund C. Stoner (1899–1968): "The Limiting Density of White Dwarf Stars," *Philosophical Magazine* **7**, 7th series, 63–70 (1929).

11. Subramanyan Chandrasekhar (1910–1995): "The Maximum Mass of Ideal White Dwarfs," *The Astrophysical Journal* **74**, 81–82 (1931).

12. Lev Landau (1908–1968): "On the Theory of Stars," *Physikalishe Zeitschrift der Sowjetunion* **1**, 285–288 (1932). Reproduced in LGSB, pp. 456–459.

13. Subramanyan Chandrasekhar (1910–1995): "The Highly Collapsed Configurations of a Stellar Mass, Second Paper," *Monthly Notices of the Royal Astronomical Society* **95**, 207–225 (1935). See MBARK, p. 336.

14. Arthur Stanley Eddington (1882–1944): "Meeting of the Royal Astronomical Society Friday, January 11, 1935," *The Observatory* **58,** 33–41 (1935), p. 38. See MBARK, pp. 336–338. Also see Arthur Stanley Eddington (1882–1944): "On 'Relativistic Degeneracy'," *Monthly Notices of the Royal Astronomical Society* **95**, 195–206 (1935), p. 195.

15. The two types of supernovae are described by Fred Hoyle (1915–2001) and William A. "Willy" Fowler (1911–1995): "Nucleosynthesis in Supernovae," *The Astrophysical Journal* **132**, 565–590 (1960); and by Virginia L. Trimble (1943–): "Supernovae. Part 1 — The Events," *Reviews of Modern Physics* **54**, 1183–1224 (1982).

16. Hans Bethe (1906–2005): "Supernova mechanisms," *Reviews of Modern Physics* **62**, 801–866 (1990).

17. Walter Baade (1893–1960) and Fritz Zwicky (1898–1974): "Cosmic Rays From Super-Novae," *Proceedings of the National Academy of Sciences* **20** (5), 259–263 (1934). Reproduced in MBARK, pp. 339–343. Also see LGSB, p. 469.

18. J. Robert Oppenheimer (1904–1967) and George M. Volkoff (1914–2000): "On Massive Neutron Cores," *Physical Review* **55**, 374–381 (1939). Reproduced in LGSB, pp. 469–477.

19. Antony Hewish (1924–): "The Diffraction of Radio Waves in Passing through a Phase-Changing Ionosphere," *Proceedings of the Royal Society of London* **A209**, 81–96 (1951); "The Diffraction of Galactic Radio Waves as a Method of Investigating the Irregular Structure of the Ionosphere," *Proceedings of the Royal Society of London* **A214**, 494–514 (1952).

20. Antony Hewish (1924–), S. Jocelyn Bell (1943–), John D. H. Pilkington (–), Paul Frederick Scott (–), and Robin Ashley Collins (–): "Observations of a Rapidly Pulsating Radio Source," *Nature* **217**, 709–713 (1968); Reproduced in LGSB, pp. 498–504, and MBARK, 513–518. We now know that the name pulsar is misleading, for the compact stars don't pulsate — they rotate, but the name has stuck. It designates a succession of repeating pulses of radio emission rather than a pulsating star.

21. John D. H. Pilkington (–), Antony Hewish (1924–), S. Jocelyn Bell (1943–), and T. W. Cole (–): "Observations of some further pulsed radio sources," *Nature* **218**, 126–129 (1968). The pulsars could probably have been discovered many years earlier, when other large radio telescopes were constructed, but radio astronomers were used to adding up signals over long time intervals to detect faint cosmic radio signals. The long time resolutions precluded the detection of the pulsars that could only be found when Hewish decided to investigate solar wind effects over short intervals of time. Pulsars are relatively faint radio sources when averaged over their period since there is no emission between the brief radio pulses.

22. Antony Hewish (1924–): "My Life in Science and Religion — A Personal Story," Seminar presented to The Faraday Institute on January 22, 2013. Also see his foreword to John Polkinghorne (1930–) and Nicholas Beale (–): *Questions of Truth*. Louisville, Kentucky: Westminster John Knox Press 2009.

23. Jocelyn Bell Burnell (1943–): *A Quaker Reflects: Can A Scientist Also Be Religious?* 2013 James Backhouse Lecture, sponsored by the Religious Society of Friends (Quakers) in Australia, pp. 3, 9, 36, 39, 41, and 43; quotation p. 39.

24. Thomas Gold (1920–2004): "Rotating Neutron Stars as the Origin of the Pulsating Radio Sources," *Nature* **218**, 731–732 (1968). Reproduced in LGSB, pp. 505–508, and MBARK pp. 518–521. But Tommy wasn't always right. Before men were sent to the Moon, he predicted that upon landing astronauts would sink into a thick layer of dust, suffocating and vanishing without a trace.

Unmanned spacecraft were set out to explore the territory, and found that there is no thick dust layer on the Moon, and that people can walk on it without sinking in over their heads.

25. David H. Staelin (1938–2011) and Edward C. Reifenstein, III (1937–): "Pulsating Radio Sources Near the Crab Nebula," *Science* **162**, 1481–1483 (1968); David W. Richards (–) and John M. Comella (–): "The Period of Pulsar NP 0532," *Nature* **222**, 551–552 (1969).

26. Herbert Friedman (1916–2000), S. W. Lichtman (–) and E. T. Byram (–): "Photon Counter Measurements of Solar X-rays and Extreme Ultraviolet Light," *Physical Review* **83**, 1025–1030 (1951).

27. Kenneth R. Lang (1941–): *Sun, Earth and Sky, Second Edition.* New York: Springer 2006, and *The Sun from Space, Second Edition.* Berlin, Heidelerg: Springer 2009.

28. Riccardo Giacconi (1931–), Herbert Gursky (1930–2006), Frank R. Paolini (–) and Bruno B. Rossi (1905–1993): "Evidence for X-rays From Sources Outside the Solar System," *Physical Review Letters* **9**, 439–443 (1962). Reproduced in LGSB, pp.62–66, and MBARK, pp. 495–202.

29. Riccardo Giacconi (1931–) *et al.*: "Discovery of Periodic X-ray Pulsations in Centaurus X-3 from *UHURU*," *The Astrophysical Journal* **167**, L67–L73 (1971).

30. John Michell (1724–1793): "On the Means of Discovering the Distance, Magnitude, etc., of the Fixed Stars, in Consequence of the Diminution of Their Light, in Case Such a Diminution Should Be Found to take Place in Any of Them, and Such Other data should be procured from observations, as would be further necessary for that purpose, *Philosophical Transactions of the Royal Society (London)* **74**, 35–57 (1784).

31. Pierre-Simon de Laplace (1749–1827): *Exposition du Systeme du Monde.* Paris: 1796. English translation by Rev. Henry H. Harte (1790–1848): Dublin: Longman Reves, Orme, Brown and Green 1830.

32. Karl Schwarzschild (1873–1916): "Über das Gravitationsfeld eines Massenpunktes nach der Einsteinschen Theorie (On the Gravitational Field of a Point Mass according to Einstein's Theory)," *Sitzungberichte der K. Preussischen Akademie der Wissenschaften zu Berlin* **1**, 189–196 (1916). [English translation in LGSB, pp. 451–455.]

33. J. Robert Oppenheimer (1904–1967) and Hartland Snyder (1913–1962): "On Continued Gravitational Contraction," *Physical Review* **56**, 455–459 (1939). Reproduced in MBARK, pp. 344–348.

34. Minoru Oda (1923–2001) *et al.*: "X-ray Pulsations From Cygnus X-1 Observed From *UHURU*," *The Astrophysical Journal* **166**, L1–L7 (1971).

35. B. Louise Webster (1941–1990) and Paul Murdin (1942–): "Cygnus X-1: A Spectroscopic Binary With A Heavy Companion?" *Nature* **235**, 37–38 (1972)

and independently Charles Thomas (Tom) Bolton (1943–): "Identification of Cygnus X-1 with HDE 226868," *Nature* **235** (2), 271–273 (1972). Reproduced in LGSB, pp. 460–465.

36. Wallace L. W. Sargent (1935–2012) *et al*.: "Dynamical Evidence for a Central Mass Concentration in the Galaxy M 87," *Astrophysical Journal* **221**, 731–744 (1978).

37. Edward E. Salpeter (1924–2008): "Accretion of Interstellar Matter by Massive Objects," *Astrophysical Journal* **140**, 796–800 (1964); Yacov Boris Zel'dovich (1914–1987): "The Fate of a Star and the Evolution of Gravitational Energy Upon Accretion," *Dokl. Akad. Nauk SSSR* **155**, 67 — March (1964), English translation in *Soviet Phys. Doklady* **9**, 195 (1964). The English astronomer James Jeans foresaw a related ejection of matter from the centers of some galaxies when he was attempting to explain why they have spiral arms. When these galaxies were still known as spiral nebulae, Jeans proposed that: "The type of conjecture which presents itself somewhat insistently is that the centers of the [spiral] nebulae are of the nature of 'singular points,' at which matter is poured into our Universe from some other, and entirely extraneous, spatial dimension, so that, to a denizen of our Universe, they appear as points at which matter is being continuously created." See James Jeans (1877–1946): *Astronomy and Cosmogony*. Cambridge, England: Cambridge University Press 1928, p. 360. Also see Martin J. Rees (1942–): "Black Hole Models for Active Galactic Nuclei," *Annual Review of Astronomy and Astrophysics* **22**, 471–506 (1984).

38. Andrea M. Ghez (1965–) *et al*.: "Stellar Orbits Around the Galactic Center Black Hole," *The Astrophysical Journal* **620** (2), 744–757 (2005); "Measuring Distance and Properties of the Milky Way's Central Supermassive Black Hole with Stellar Orbits," *The Astrophysical Journal* **689** (2), 1044–1062 (2008).

39. Fiction writers have presented captivating allusions to whirling masses that swallow up everything near them. Edgar Allan Poe (1809–1949) described one of them in his short story *A Decent into the Maelström*, where a ship and most of its crew were caught in "the most terrible hurricane that ever came out of the heavens," and pulled down into the whirling sea. In Hermann Melville's (1819–1891): *Moby Dick*, the whaling ship Pequod and most of its crew also drown within spinning circles that carry them down and out of sight.

12. Darkness Made Visible

1. William Shakespeare (1564–1616), *Henry V: Act IV, Prologue, Chorus* (1599), lines 1 to 3.

2. *The Holy Bible, King James Version, The Book of Psalms, Psalm 18*: 11.

3. The American poet Henry Wadsworth Longfellow (1807–1882) wrote a *Hymn to the Night*. The lyrics of *The Music of the Night* were written by Charles Hart (1961–) and sung in *The Phantom of the Opera*, 1986.

4. Edward Emerson Barnard (1857–1923): "On the Dark Markings of the Sky, with a Catalogue of 182 Such Objects," *Astrophysical Journal* **49**, 1–24 (1919); *A Photographic Atlas of Regions of the Milky Way*, Washington, D.C., Carnegie Institution 1927. See MBARK, pp. 398–403.

5. Maximilian Franz Joseph Cornelius Wolf (1863–1932): "Über den dunklen Nebel NGC 6960 (On the Dark Nebula NGC 6960)," *Astronomische Nachrichten* **219**, 109–116 (1923). English translation in LGSB, pp. 566–571; Robert J. Trumpler (1886–1956): "Preliminary Results on the Distances, Dimensions, and Space Distribution of Open Star Clusters," *Lick Observatory Bulletin 14*, no. 420, 154–188 (1930). Reproduced in LGSB, pp. 593–604, and MBARK, pp. 403–406, Hendrik C. van de Hulst (1918–2000): "The Solid Particles of Interstellar Space," *Recherches astronomiques de l'Oservatoire d'Utrecht* **11**, pt. 2, 1–50 (1949). Reproduced in LGSB pp. 605–611.

6. Lao-tzu (571 BC–531 BC): *Tao Te Ching* (6th century BC).

7. T. S. Eliot (1888–1965): *Four Quartets, "East Coker," III*, lines 1, 2. New York: Harcourt 1943.

8. Rabindranath Tagore (1861–1941): *Fireflies*. New York: Macmillan 1933, p. 267.

9. Kenneth R. Lang (1941–): "Serendipitous Astronomy," *Science* **327**, 29–40 (2010).

10. Heinrich Hertz (1857–1894): "Ueber die Ausbreitungsgeschwindigkeit der electrodynamischen Wirkungen," *Annalen der Physik* **270**, 551–569 (1888); *Untersuchungen über die Ausbreitung der elektrischen Kraft* (1893), English translation by Daniel Evan Jones (1860–1941): *Electric Waves: Being Researches On the Propagation of Electric Action with Finite Velocity Through Space*. London and York: Macmillan and Co. 1893.

11. Guglielmo Marconi (1874–1937): "Wireless Telegraphic Communication," Nobel Lecture, December 11, 1909.

12. Karl G. Jansky (1905–1950): "A Note on the Source of Interstellar Interference," *Proceedings of the Institute of Radio Engineers* **23**, 1158–1163 (1935). Reproduced in LGSB, pp. 30–33, and MBARK, pp. 455–462.

13. Grote Reber (1911–2002): "Cosmic Static," *Astrophysical Journal* **100**, 279–287 (1944). Reproduced in LGSB, pp. 30–35, and MBARK, pp. 462–464.

14. In 1912, the ardent amateur balloonist Victor Franz Hess found that energetic ionizing radiation increased in intensity high in the Earth's atmosphere and appeared to be coming from above it. Caltech physicist Robert Millikan and his students used numerous unmanned balloon flights, as well as measurements atop mountains and deep underwater, to confirm that the "radiation" was coming

from outside our atmosphere and incidentally gave the radiation the name *cosmic rays*. Millikan thought the cosmic rays were associated with the "birth pangs of creation," but they are now attributed to acceleration during explosive stellar death. They were subsequently found to be charged particles deflected by the Earth's magnetic field rather than "rays." See Victor Franz Hess (1883–1964): "Concerning Observations of Penetrating Radiation on Seven Free Balloon Flights," *Physikalische Zeitschrift* **13**, 1084–1091 (1912). Reproduced in LGSB, pp. 13–10, and MBARK, pp. 279–284. Robert A. Millikan (1868–1953) and G. H. Cameron (–): "High Frequency Rays of Cosmic Origin. III — Measurements in Snow-Fed Lakes at High Altitudes," *Physical Review* **28**, 851–868 (1926); Robert A. Millikan (1868–1953) and G. H. Cameron (–): "The Origin of Cosmic Rays," *Physical Review* **32**, 533–537 (1928).

15. Hendrik C. Van De Hulst (1918–2000): "Radio Waves from Space: Origin of Radio Waves," *Nederlands tijdschrift voor Natuurkunde* **11**, 210–221 (1945). [English translation in LGSB, pp. 627–632, and partially reproduced in MBARK, pp. 467–468.]

16. Iosif S. Shkloviskii (1916–1985): "On the Nature of Galactic Radio Emission — in Russian," *Astronomicheskii Zhurnal* **29**, 418–449 (1952).

17. Harold I. Ewen (1922–) and Edward M. Purcell (1912–1997): "The Radio Frequency Detection of Interstellar Hydrogen," *Nature* **168**, 356 (1951) Reproduced in LGSB, pp. 633–635, and MBARK, pp. 469–470.

18. C. Alex Muller (–) and Jan H. Oort (1900–1992): "The Interstellar Hydrogen Line at 1,420 MHz and an Estimate of Galactic Rotation," *Nature* **168**, 356–358 (1951). Reproduced in LGSB, pp. 633–637. The Princeton astronomer Lyman Spitzer had recently inferred a similar temperature when considering the various heating and cooling processes of interstellar space. See Lyman Spitzer (1914–1997) and Malcolm P. Savedoff (1928–): "The Temperature of Interstellar Matter," *Astrophysical Journal* **111**, 593–608 (1950). Reproduced in LGSB, pp. 617–626.

19. Jan H. Oort (1900–1992), Frank J. Kerr (1918–2000), and Gart Westerhout (1927–2012): "The Galactic System as a Spiral Nebula," *Monthly Notices of the Royal Astronomical Society* **118**, 379–389 (1958). Reproduced in LGSB, pp. 643–651, and MBARK, pp. 433–439.

20. Charles H. Townes (1915–2015): *How the Laser Happened: Adventures of a Scientist*. Oxford: Oxford University Press 1999, p. 30.

21. Charles H. Townes (1915–2015): *Ibid.* pp. 65, 190–191.

22. Arthur L. Schawlow (1921–1999) and Charles H. Townes (1915–2015): "Infrared and Optical Masers," *Physical Review* **112**, 1940–1949 (1958).

23. Charles H. Townes (1915–2015): "Microwave and Radiofrequency Resonance Lines of Interest to Radio Astronomy," *Radio Astronomy: Proceedings of the International Astronomical Union Symposium Number 4.* Ed. Hendrik C. Van De Hulst (1918–2000). Cambridge, England: Cambridge University Press 1957, pp. 92–103.

24. Sander Weinreb (1936–), Alan H. Barrett (1927–1991), Marion Littleton "Lit" Meeks (1923–2013), and John C. Henry (–): "Radio Observations of OH in the Interstellar Medium," *Nature* **200**, 829–831 (1964). Reproduced in LGSB, pp. 666–670, and MBARK, pp. 470–475.

25. Charles H. Townes (1915–2015): "The Convergence of Science and Religion," *Think* **32**, No. 2, 2–7 (1966). For the accidental and unexpected nature of scientific discovery also see Kenneth R. Lang (1941–): "Serendipitous Astronomy," *Science* **327**, 39–40 (2010).

26. Al Cheung (–) *et al.*: "Detection of NH_3 Molecules in the Interstellar Medium by Their Microwave Emission," *Physical Review Letters* **21**, 1701–1705 (1968); S. H. Knowles (–) *et al.*: "Spectra, Variability, Size and Polarization of H_2O Microwave Emission Sources in the Galaxy," *Science* **163**, 1055–157 (1969).

27. L. E. Snyder (–), D. Buhl (–), Benjamin Zuckerman (1943–), and Patrick E. Palmer (–): "Microwave Detection of Interstellar Formaldehyde," *Physical Review Letters* **22**, 679– (1969); Robert W. Wilson (1936–), K. B. Jefferts (–), and Arno A. Penzias (1933–): "Carbon Monoxide in the Orion Nebula," *The Astrophysical Journal* **161**, L43–L44 (1970).

28. Agnes M. Clerke (1842–1907): *Problems in Astrophysics.* London: Adam and Charles Black 1903, p. 400.

29. Fritz Zwicky (1898–1974): "On the Masses of Nebulae and of Clusters of Nebulae," *Astrophysical Journal* **86**, 217–246 (1937). Reproduced in LGSB, pp. 729–737, and MBARK, pp. 559–564.

30. Fritz Zwicky (1898–1974): "Die Rotverschiebung von extragalaktischen Nebeln," *Helvetica Physica Acta* **6**, 110 (1933). [English translation by Sydney van den Bergh (1929–) in "The Early History of Dark Matter," *Publications of the Astronomical Society of the Pacific* **111**, 657 (1999).]

31. Fritz Zwicky (1898–1974): "Nebulae as Gravitational Lenses," *Physical Review* **51**, 290 (1937).

32. C. Roger Lynds (1928–) and Vahe Petrosian (1938–): "Luminous arcs in clusters of galaxies," *Astrophysical Journal* **336**, 1–8 (1989).

33. Nohubiro Okabe (–) *et al.* : "LoCuSS : The Mass Density Profile of Massive Galaxy Clusters at z = 0.2," *Astrophysical Journal Letters* **769**, L35–L41 (2013).

34. Vera C. Rubin (1928–2016) and W. Kent Ford, Jr. (1931–): "Rotation of the Andromeda Nebula from a Spectroscopic Survey of Emission Regions," *The Astrophysical Journal* **150**, 379–403 (1970).

35. Morton S. Roberts (–) and Arnold H. Rots (–): "Comparison of Rotation Curves of Different Galaxy Types," *Astronomy and Astrophysics* **26**, 483–485 (1973).

36. Vera C. Rubin (1928–2016), W. Kent Ford, Jr. (1931–), and Norbert Thonnard (–): "Rotational Properties of 21 Sc Galaxies with a Large Range of Luminosities and Radii. From NGC 4605 (R = 4 kpc) to NCG 2885 (R = 122 kpc)," *Astrophysical Journal* **238**, 471–487 (1980), Reproduced in MBARK, pp. 559–567.

37. Jeremiah P. Ostriker (1937–), P. James E. Peebles (1935–), and Amos Yahil (–):" The Size and Mass of Galaxies, and the Mass of the Universe," *Astrophysical Journal Letters* **193**, L1–L4 (1974).

38. Vera C. Rubin (1928–2016), in *Origins: The Lives and Worlds of Modern Cosmologists* [Ed. Alan Lightman (1948–) and Roberta Brawer (–)], Cambridge, Massachusetts: Harvard University Press 1990, pp. 296, 303. Also see Vera C. Rubin (1928–2016): *Bright Galaxies, Dark Matters*. Woodbury, New York: American Institute of Physics Press 1997, p. 219.

39. C. J. Copi (–), David N. Schramn (1945–1997), and Michael S. Turner (1949–): "Big-Bang Nucleosynthesis and the Baryon Density of the Universe," *Science* **267**, 192–199 (1995).

40. *WMAP* Astronomers: "Five-Year Wilkinson Microwave Anisotropy Probe Observations," *Astrophysical Journal Supplement* **180**, Issue 2 (2009), Seven-Year Wilkinson Microwave Anisotropy Probe Observations, *Astrophysical Journal Supplement* **192**, Issue 2 (2011).

41. *Planck* Collaboration : "*Planck* 2013 Results," *Astronomy and Astrophysics* **1303**, thirty four articles (2014). Also see *Planck* 2015 Result.

13. Primordial Light

1. *The Holy Bible, King James Version, The First Book of Moses, called Genesis 1*: 1–5.

2. Albert Einstein (1879–1955): "Kosmologische Betrachtungen zur allgemeinen Relativitätstheorie (Cosmological Considerations of the General Theory of Relativity)," *Sitzungsberichte de Preußischen Akademie der Wissenschaften zu Berlin* **1**, 142–152 (1917). English translation in: *The Principle of Relativity* (Ed. A. Sommerfeld), New York: Dover 1952, pp. 175–188. See MBARK, pp. 315–316.

3. Willem de Sitter (1872–1934): "On Einstein's Theory of Gravitation, and Its Astronomical Consequences. Third Paper," *Monthly Notices of the Royal Astronomical Society* **78**, 3–28 (1917). See MBARK, p. 316.

4. Georges Lemaître (1894–1966): "Un univers homogéne de masse constante et de rayon croissant rendant compte de la vitesse radiale des nébuleuses extra-galactiques," *Annales de la Société scientifique de Bruxelles* **A47**, 49–56 (1927).

5. Georges Lemaître (1894–1966): "A Homogeneous Universe of Constant Mass and Increasing Radius accounting for the Radial Velocity of Extra-Galactic Nebulae," *Monthly Notices of the Royal Astronomical Society* **91**, 483–490 (1931). Reproduced in LGSB, pp. 844–848, quotation p. 848, and MBARK, pp. 317–319.

6. Georges Lemaître (1894–1966): "L'expansion de l'espace," *La Revue des Questions Scientifiques, 4e, Série*, November **17**, 391–400 (1931). Lemaître was unaware of earlier work by the Russian mathematician Aleksandr Alexsandrovich Friedmann who in 1922 first considered dynamical, non-stationary solutions to Einstein's theory for the entire Cosmos. Friedmann found contracting, expanding and oscillating model Universes, but failed to suggest any physical significance to his equations or any connection between them and astronomical observations. After taking a balloon to record-breaking heights, for meteorological and medical purposes, Friedmann became ill and died in 1925 at the relatively young age of 37, and his papers remained largely unnoticed and unread. See Alexsandrovich Friedmann (1888–1925): "Über die Krümmung des Raumes (On the Curvature of Space)," *Zeitschrift für Physik* **10**, 377–386 (1922). Reproduced in English translation in LGSB, pp. 838–843.

7. Georges Lemaître (1894–1966): "The Beginning of the World From the Point of View of Quantum Theory," *Nature* **127**, No. 3210, 706 (1931). Reproduced in MBARK, pp. 322–324. In 1848, almost a century before Lemaitre's proposals, the American poet and novelist Edgar Allan Poe (1809–1849) wrote *Eureka: A Prose Poem*, which also supposed that the Universe arose from the explosion of a simple, primordial particle in an instantaneous flash.

8. Georges Lemaître (1894–1966): *L'Hypothèse de l'Atome Primitif: Essai de Cosmogonie*. Paris: Neuchatel, Éditions du Griffon 1946. English translation as *The Primeval Atom: An Essay on Cosmogony*, New York: D. Van Nostrand Co. 1950, p. 78.

9. Albert Einstein (1879–1955): "Kosmologische Betrachtungen zur allgemeinen Relativitätstheroie (Cosmological Considerations of the General Theory of Relativity)," *Koniglich Preussische Academie der Wissenshaften, Sitzungsberichte (Berlin)* **1**, 142–152 (1917). [English translation in *The Principle of Relativity*, Ed. Arnold Sommerfeld (1868–1951), New York: Dover 1952.] Albert Einstein (1879–1955) and Willem de Sitter (1872–1934): "On the Relation Between the Expansion and Mean Density of the Universe," *Proceedings of the National Academy of Sciences* **18**, 213–214 (1932). Reproduced in LGSB, pp. 849, 850, and MBARK, pp. 313–317.

10. Arthur Stanley Eddington (1882–1944): *The Nature of the Physical World*. Cambridge, England: Cambridge University Press 1928, p. 85; "The End of the World from the Standpoint of Mathematical Physics," *Nature* **127**, 447–453 (1931), p. 450; *The Expanding Universe*. Cambridge, England: Cambridge University Press 1933, pp. 55 and 56. The italics in the quotation are Eddington's.

11. Arthur Stanley Eddington (1882–1944): *New Pathways in Science*. Cambridge, England: Cambridge University Press 1934, p. 220. The parenthetical reference to Lemaître is Eddington's.

12. Hermann Bondi (1919–2005) and Thomas Gold (1920–2004): "The Steady-State Theory of the Expanding Universe," *Monthly Notices of the Royal Astronomical Society* **108**, 252–270 (1948). Reproduced in LGSB, pp. 853–863, and MBARK, pp. 324–326.

13. Fred Hoyle (1915–2001): "A New Model for the Expanding Universe," *Monthly Notices of the Royal Astronomical Society* **108**, 372–382 (1948).

14. Georges Lemaître (1894–1966): "Un univers homogéne de masse constante et de rayon croissant rendant compte de la vitesse radiale des nébuleuses extra-galactiques," *Annales de la Société scientifique de Bruxelles* **A47**, 49–56 (1927); "A Homogeneous Universe of Constant Mass and Increasing Radius accounting for the Radial Velocity of Extra-Galactic Nebulae," *Monthly Notices of the Royal Astronomical Society* **91**, 483–490 (1931). Reproduced in LGSB, pp. 844–848, and MBARK, pp. 317–319.

15. George Gamow (1904–1968): *The Creation of the Universe*. London: MacMillan & Co. 1952, p. 42.

16. Ralph A. Alpher (1921–2007) and Robert C. Herman (1914–1997): "Evolution of the Universe," *Nature* **162**, 774–775 (1948), Reproduced in MBARK, pp. 363–365; Ralph A. Alpher (1921–2007), J. W. Follin, Jr. (–), and Robert C. Herman (1914–1997): "Physical Conditions in the Initial Stages of the Expanding Universe," *Physical Review* **92**, 1347–1361(1953).

17. Helge Kragh (1944–): *Cosmology and Controversy: The Historical Development of Two Theories of the Universe*. Princeton, New Jersey: Princeton University Press 1996.

18. George Gamow (1904–1968): *American Institute of Physics, Oral History Transcript of Interview with George Gamow* by Charles Weiner (1932–2012) on April 25, 1968, p. 2.

19. George Gamow (1904–1968): "The Quantum Theory of Nuclear Disintegration," *Nature* **122**, 805–806 (1928); "Zur Quantentheorie der Atomzertrümmerung (On the Quantum Theory of the Atomic Nucleus)," *Zeitschrift für Physik* **52**, 510–515 (1928).

20. George Gamow (1904–1968): *American Institute of Physics, Oral History Transcript of Interview with George Gamow* by Charles Weiner (1932–2012) on April 25, 1968, p. 105.

21. George Gamow (1904–1968): *Mr. Tompkins in Wonderland.* Cambridge, England: Cambridge University Press 1940; *Mr. Tompkins Explores the Atom.* Cambridge, England: Cambridge University Press 1944; *One, Two, Three … Infinity: Facts and Speculations of Science.* New York: Viking Press 1947, revised 1961; *Thirty Years that Shook Physics: The Story of Quantum Theory.* Garden City, New Jersey: Doubleday and Company 1966, reproduced New York: Dover Publications.

22. Karl Hufbauer (–); *George Gamow 1904–1908: A Biographical Memoir*, Washington, D.C.: National Academy of Sciences 2009, pp. 29, 30.

23. Simon Mitton (1946–): *Fred Hoyle: A Life in Science.* Cambridge, England: Cambridge University Press 2011, p. 22.

24. Fred Hoyle (1915–2001): The derogatory expression *Big Bang* was given during a radio broadcast that was published in *The Listener* **41**, 567 (1949), and reproduced in *The Nature of the Universe.* Oxford: Blackwell 1953, p. 94. The sarcastic term *Bang* was previously coined for the explosive beginning of the expanding Universe by Arthur Stanley Eddington (1882–1944): *The Expanding Universe.* Cambridge, England: Cambridge University Press 1933, p. 56; *New Pathways in Science.* Cambridge, England: Cambridge University Press 1934, p. 220.

25. Simon Mitton (1946–): *Conflict in the Cosmos: Fred Hoyle's Life in Science.* Washington, D. C.: Joseph Henry Press 2005.

26. Fred Hoyle (1915–2001): *The Nature of the Universe.* Oxford: Basil Blackwell 1953, p. 111.

27. D. Dunbar (–), F. Noel (–), Ralph E. Pixley (–), William A. Wenzel (–), and Ward Whaling (1923–): "The 7.68-MeV state in C^{12}," *Physical Review* **92**, 649, 650 (1953).

28. Fred Hoyle (1915–2001): In Mervyn Stockwood (Ed., 1913–1995): *Religion and the Scientists.* London: SCM Press 1959, p. 64.

29. Fred Hoyle (1915–2001): "The Universe: Some Past and Present Reflections," *Engineering and Science (California Institute of Technology Alumni Magazine)* **45**, 2 (November), 8–12 (1981), p. 12.

30. Fred Hoyle (1915–2001): *Home is Where the Wind Blows: Chapters from a Cosmologist's Life.* Mill Valley, California: University Science Books 1994, p. 259.

31. Fred Hoyle (1915–2001): "Recent Developments in Cosmology," *Nature* **208**, 111–114 (1965). Although Hoyle withdrew his support for the Steady State Cosmology in this article, he remained skeptical of the Big Bang theory

throughout his life because it required a unique moment in space and time, with insufficient capacity, he thought, for the development of biological systems without Divine adjustments.

32. Arno A. Penzias (1933–) and Robert W. Wilson (1936–): "A Measurement of Excess Antenna Temperature at 4080 MHz," *Astrophysical Journal* **142**, 419–421 (1965). Reproduced in LGSB, pp. 873–876, and MBARK, pp. 508–512.

33. Robert H. Dicke (1916–1997), P. James E. Peebles (1935–), Peter G. Roll (1935–), and David T. Wilkinson (1935–2002): "Cosmic Black Body Radiation," *Astrophysical Journal* **142**, 414–419 (1965).

34. Robinson Jeffers (1887–1962): *The Great Explosion*, lines 8 to 16, in *The Beginning and the End and other Poems*. New York: Random House 1963, p. 3.

35. Robert Wilson (1933–): Interview in 1982, reported by Jeremy Bernstein (1929–) in *Three Degrees Above Zero*, New York, Scribner's 1984, page 205.

36. John C. Mather (1946–) and John Boslough (1942–2010): *The Very First Light: The True Inside Story of the Scientific Journey Back to the Dawn of the Universe.* New York: Basic Books, Harper Collins Publishers 1996.

37. John C. Mather (1946–) *et al.*: "Measurement of the Cosmic Microwave Background Spectrum by the *COBE* FIRAS Instrument," *Astrophysical Journal* **420**, 439–512 (1994). Also see John C. Mather (1946–) and John Boslough (1942–2010): *Ibid.* pp. 229, 235.

38. George F. Smoot (1945–) *et al.*: "Structure in the *COBE* Differential Microwave Radiometer First-Year Maps," *Astrophysical Journal Letters* **396**, L1–L5 (1992), "Summary of results from COBE," *AIP Conference Proceedings* **476**, 1–10 (1999).

39. Charles L. Bennett (1956–) *et al.*: "First-year Microwave Anisotropy Probe (*WMAP*) preliminary maps and basic results." *Astrophysical Journal Supplement* **248**, Issue 2, 2–27 (2003). These results have been amplified and extended using instruments aboard the *Planck* spacecraft, launched on May 14, 2009 and operated by the European Space Agency. The results announced in 2013 indicate that the Universe contains about 5% ordinary matter, 26% dark matter, and 69% dark energy. See *Planck* Collaboration: "Planck 2013 Results," *Astronomy and Astrophysics* **1303** (2014).

14. The Origin of the Chemical Elements

1. Benjamin "Ben" Jonson (1572–1637): *The Alchemist*. London: Printed by Thomas Snodham for Walter Burre 1612. Subtle, the Alchemist, is speaking in *Act 2, Scene 2.1*.

2. John Dalton (1766–1844): "On the Absorption of Gases by Water and Other Liquids," *Memoirs of the Literary and Philosophical Society of Manchester* 1803;

A New System of Chemical Philosophy. Manchester, London: R. Bickerstaff, Strand 1808, 1810, 1827.

3. Johann Jakob Balmer (1825–1898): "Notiz uber die Spectrallinien des Wasseroffs (Note on the Spectral Lines of Hydrogen)," *Annalen der Physik und Chemie* **25**, 80–85 (1885).

4. Gustav Kirchhoff (1824–1887) and Robert Bunsen (1811–1899): "Chemische Analyse durch Spektralbeobachtungeii (Chemical Analysis by Observation of Spectra)," *Annalen der Physik und der Chemie* **110**, 161–189 (1860).

5. Meghnad Saha (1893–1956): "Ionization in the Solar Chromosphere," *Philosophical Magazine* **40**, 479–488 (1920). Reproduced in LGSB, pp. 236–242.

6. William Wordsworth (1770–1850): *Lines Composed a Few Miles above Tintern Abbey, on Revisiting the Banks of the Wye during a Tour, July 13, 1798*, Stanza 6, lines 122, 123. Thomas H. Huxley (1825–1895): *Letter to Charles Kingsley* on September 23, 1860, in Leonard Huxley (Ed. 1860–1933), *Letters of Thomas Henry Huxley, Volume 1.* New York: D. Appleton and Co. 1900, p, 235.

7. Cecilia H. Payne (1900–1979): *Stellar Atmospheres*, Cambridge, England: W. Heffer and Sons 1925, *Harvard Observatory Monograph No. 1.* The Chapter on "The Relative Abundances of the Elements," is reproduced in LGSB, pp. 243–248, and MBARK, pp. 250–256.

8. Henry Norris Russell (1877–1957): "On the Composition of the Sun's Atmosphere" *Astrophysical Journal* **70**, 11–82 (1929); *The Composition of the Stars, Halley Lecture.* Oxford: Oxford University Press 1933. In this paper Russell also provided values for the relative abundances of hydrogen, carbon, nitrogen, oxygen, silicon and iron that are within a factor of 2 of modern determinations.

9. Bengt Strömgren (1908–1987): "The Opacity of Stellar Matter and the Hydrogen Content of the Stars," *Zeitschrift für Astrophysik* **4**, 118–152 (1932). Bengt Strömgren (1908–1987): "On the Interpretation of the Hertzsprung-Russell-Diagram," *Zeitschrift für Astrophysik* **7**, 222–259 (1933).

10. William Draper Harkins (1873–1951): "The Evolution of the Elements and the Stability of Complex Atoms," *Journal of the American Chemical Society* **39**, 856–879 (1917).

11. Arthur Stanley Eddington (1882–1944): "The Internal Constitution of the Stars," *Nature* **106**, 14–20 (1920); *Observatory* **43**, 341–358 (1920); Reproduced in LGSB, pp. 281–290. Also see Eddington's book: *The Internal Constitution of the Stars.* Cambridge, England: Cambridge University Press 1926.

12. Hans E. Suess (1909–1993) and Harold C. Urey (1893–1981): "Abundances of the Elements," *Reviews of Modern Physics* **28**, 53–74 (1956).

13. George Gamow (1904–1968): "Expanding Universe and the Origin of the Elements," *Physical Review* **70**, 572–573 (1946).

14. Ralph A. Alpher (1921–2007), Hans Bethe (1906–2005) and George Gamow (1904–1968): "The Origin of Chemical Elements," *Physical Review* **73**, 803–804 (1948). Reproduced in LGSB, pp. 864–865, and MBARK, pp. 358–362. Two years after this α-β-γ paper, Chushiro Hayashi (1920–2010) of the Nanikawa University in Japan showed that in the first moments of the expansion the temperature was hot enough to create more exotic particles, such as positrons, the anti-matter particles of the electrons. See Chushiro Hayashi (1920–2010): "Proton-Neutron Concentration Ratio in the Expanding Universe at the Stages Preceding the Formation of the Elements," *Progress in Theoretical Physics (Japan)* **5**, 224–235 (1950).

15. Robert V. Wagoner (1938–), William A. Fowler (1911–1995), and Fred Hoyle (1915–2001): "Nucleosynthesis in the Early Stages of the Expanding Universe," *Science* **152**, 677–678 (1966); "On the Synthesis of Elements at Very High Temperature," *Astrophysical Journal* **148**, 3–49 (1967).

16. Hans A. Bethe (1906–2005): "Energy Production in Stars," *Physical Review* **55**, 434–456 (1939), Reproduced in LGSB, pp. 320–338, and MBARK, pp. 348–357; Carl Friedrich von Weizsäcker (1912–2007): "Über Elementumwandlungen in Innern der Sterne II (Element Transformation Inside Stars II)," *Physikalische Zeitschrift* **39**, 633–646 (1938). [English translation in LGSB, pp. 309–319.]

17. Ernst J. Öpik (1893–1985): "Stellar Models with Variable Compositions II. Sequences of Models with Energy Generation Proportional to the 15th Power of Temperature," *Proceedings of the Royal Irish Academy, Section A,* **54**, 49–57 (1951), *Contributions from the Armagh Observatory,* no. 3 (1951). Edwin E. Salpeter (1924–2008): "Nuclear Reactions in Stars without Hydrogen," *Astrophysical Journal* **115**, 326–328 (1952). Reproduced in LGSB, pp. 349–352.

18. Fred Hoyle (1915–2001): "The Synthesis of the Elements from Hydrogen," *Monthly Notices of the Royal Astronomical Society* **106**, 343–383 (1946); "On Nuclear Reactions Occurring in Very Hot Stars. I. The Synthesis of Elements from Carbon to Nickel," *Astrophysical Journal Supplement* **1**, 121–146 (1954); Fred Hoyle (1915–2001) and William A. Fowler (1911–1995): "Nucleosynthesis in Supernovae," *Astrophysical Journal* **132,** 565–590 (1956).

19. William A. Fowler (1911–1995), Georgeanne R. Caughlan (1916–1994) and Barbara A. Zimmerman (–): "Thermonuclear Reaction Rates I, II," *Annual Reviews of Astronomy and Astrophysics* **5**, 525–569 (1967), **13**, 69–111 (1975), **21**, 165–176 (1983).

20. E. Margaret Burbidge (1919–2010), Geoffrey R. Burbidge (1925–2010), William A. Fowler (1911–1995), and Fred Hoyle (1915–2001): "Synthesis of

the Elements in Stars," *Reviews of Modern Physics* **29**, 547–650 (1957). Reproduced in LGSB, pp. 374–388, and MBARK, pp. 366–376.

21. Walt Whitman (1819–1892): *Leaves of Grass. Book III Song of Myself* (1892), Section 31, line 1.

22. Joni Mitchell (1943–): *Woodstock Lyrics*, August 1969, lines 9, 10.

Author Index

Subject Index